Oxford Science Research Papers 4

Ab Initio Molecular Orbital Calculations for Chemists

W. G. Richards
J. A. Horsley

CLARENDON PRESS·OXFORD·1970

Oxford University Press, Ely House, London W.1.

GLASGOW NEW YORK TORONTO MELBOURNE WELLINGTON
CAPE TOWN SALISBURY IBADAN NAIROBI DAR ES SALAAM LUSAKA
ADDIS ABABA BOMBAY CALCUTTA MADRAS KARACHI LAHORE DACCA
KUALA LUMPUR SINGAPORE HONG KONG TOKYO

*Set in cold type by E.W.C. Wilkins & Associates Ltd., London
and Printed in Great Britain by
S. Glossop and Sons Ltd., Cardiff*

CONTENTS

INTRODUCTION

Practical chemists and molecular physicists are already familiar with semi-empirical molecular orbital calculations. Simple Hückel approximations in organic chemistry and Wolfsberg–Helmholtz calculations in inorganic chemistry have enabled the non-theoretical chemist to interpret his experiments and to estimate unknown quantities not amenable to direct measurement.

Ab initio calculations, which do not involve the rather drastic approximations of the simpler methods, have until recently been the province of pure theoreticians, and the method has been thought too complicated to be worth the trouble of a non-specialist. This situation has now changed. The massive computer programmes developed for this type of work have evolved to the stage where no molecule is in principle beyond their capabilities. Furthermore, some of the programmes have been made freely available by their authors, so that an organic chemist or a spectroscopist can obtain an *ab initio* wave function programme, which he may use as a black box rather as he might use a spectrograph. The analogy of an *ab initio* wave function programme with a spectrograph is fairly precise. Although the spectroscopist uses a commercial instrument he nonetheless requires much background knowledge and skill. Just as there is no point in taking spectra for their own sake, so merely computing a molecular wave function is useless unless it is done with care and for a good reason. Furthermore, unless the computation is done with care 'impurities' can ruin wave functions. The interpretation of the results in both cases depends on the skill of the practitioner. This book attempts to give the non-specialist the knowledge necessary to do this.

The book is aimed directly at the practical chemist to cover *ab initio* calculations rather as the book of Roberts (*Notes on molecular orbital calculations*) covered the simple linear-combination-of-atomic-orbital (LCAO) molecular-orbital technique.

That these large *ab initio* calculations are moving into the realm of the non-theoretical scientist is evidenced by a wealth of recent literature, and the continued development of computers is only going to accelerate this process. Although much more rigorous and hence more complicated than the approximate methods, such is the speed of modern computers that *ab initio* wave functions may be obtained in a matter of seconds for a diatomic molecule and in a few minutes for a small polyatomic molecule.

In this book the necessary background knowledge will be given, but as little mathematics as possible is included and proofs will be omitted, since they may be found in any of the large number of books on quantum chemistry. It is assumed that the reader has followed a course in elementary quantum chemistry, including applications of group theory. From the experience of the authors in teaching these techniques to graduate students it is clear that chemists can understand quantum mechanics most quickly when given a specific example to illustrate manipulations. The procedures are often clearer when going from the particular case to the general than vice versa. As illustrations the examples of BH, CO, N_2, H_2O, NH_3, and MF_6 will be taken. These small molecules have been chosen since they have sufficiently few electrons to keep formulae down to a manageable length but are nonetheless large enough to include all the areas of difficulty to be found in bigger molecules.

INTRODUCTORY SUMMARY OF QUANTUM MECHANICS

This short chapter makes no attempt at rigour or completeness. It merely tries to indicate what the relevant parts of quantum mechanics are when dealing with molecular calculations. The detailed theory is covered in a number of excellent textbooks. [1-5]

1.1. Wave functions

The beautifully simple Schrödinger equation in the form

$$H\psi = E\psi,$$

although elegant to the specialist, hides a great deal from the non-theoretician by its very simplicity. It merely states that any system, (in our particular case an isolated molecule) is represented by ψ, its wave function, which is in general complex. When an operator consisting of kinetic and potential energy terms acts upon this function we have a differential eigenvalue equation, the eigenvalue being the energy of the system.

Taking this equation $H\psi = E\psi$, multiplying both sides by ψ^*, the complex conjugate, and integrating over all coordinates, we can obtain the formula

$$E = \frac{\int \psi^* H \psi \, d\tau}{\int \psi^* \psi \, d\tau}$$

An important theorem known as the variation principle tells us that the closer our wave function ψ comes to the true wave function the lower will be the value of the energy obtained from the above equation. (This is true only for the lowest state of a given symmetry.)

An alternative and much-used notation for this equation is

$$E = \frac{<\psi \,|\, H \,|\, \psi>}{<\psi \,|\, \psi>}$$

where $<\psi| \equiv \psi^*$ and $|\psi> \equiv \psi$, and integration over all space is also assumed.

What is ψ? A wave function is just a mathematical function like any other.

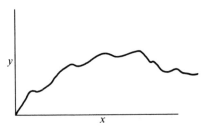

In quantum mechanics we postulate that ψ must be a well-behaved quadratically-integrable function that goes to zero at infinity for a bound state.

It can be operated on by an appropriate operator to give a value of any physical observable of the system as an eigenvalue of the operator equation (as long as the observable is a constant of motion). There is an operator for each observable, but we will restrict our attention to the energy.

In the case of the hydrogen atom the function that describes the properties of the system is obviously a three-dimensional one and can be represented in polar coordinates as a product of radial (r-dependent) and angular (θ- and ϕ-dependent) terms:

$$\psi = R(r)\,\Theta(\theta)\,\Phi(\phi).$$

The well-known drawings of hydrogen-atom wave functions or orbitals, e.g.

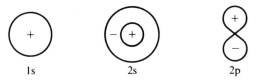

are merely two-dimensional representations of the three-dimensional wave function. The signs refer to the radial part, i.e.

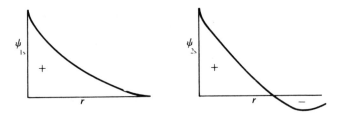

The wave function ψ should not be confused with $\psi^*\psi$. This is a measure of probability or electron density for a one-electron wave function, and has no negative areas on the three-dimensional representations so commonly drawn in elementary chemistry books.

For a molecule with many electrons we make an approximation known as the *orbital approximation*. We assume that each electron has its separate wave function or orbital. This orbital is an eigenfunction of a one-electron Hamiltonian operator H.

An orbital is then just an ordinary mathematical function describing the properties of one electron in a molecule. It is a three-dimensional function and for convenience is expressed in polar coordinates, when expanded in terms of atomic orbitals.

1.2. Antisymmetric wave functions

If we include spin in our wave functions we have *spin orbitals*, the spatial part being multiplied by a spin function α or β, and again we would have our molecular or atomic wave function as a product of spin orbitals, one for each electron.

Let us take as an example the lithium atom. The structure is $1s^2 2s$, or the wave function

$$\psi = \phi_{1s}^{\alpha} \, \phi_{1s}^{\beta} \, \phi_{2s}^{\alpha} \, ,$$

or, in a conventional notation,

$$\psi = \phi_{1s} (1) \, \bar{\phi}_{1s} (2) \, \phi_{2s} (3),$$

where the 1, 2, and 3 in brackets refer to the three electrons. A suitable ϕ_{1s} will be of the form

$$\frac{1}{\sqrt{\pi}} \left(\frac{Z}{a_0} \right)^{3/2} \exp \left(-\frac{Z}{2a_0} r \right)$$

where Z is the nuclear charge, ϕ_{2s} is similarly

$$\frac{1}{\sqrt{(32\pi)}} \left(\frac{Z}{a_0} \right)^{3/2} \left(2 - \frac{Z}{a_0} r \right) \exp \left(-\frac{Z}{2a_0} r \right).$$

This idea of the wave function being a simple product ignores the Pauli Principle, which states that the total wave function should be antisymmetric with respect to electron permutation. Hence in the above case if we changed electrons (1) and (2) we would not have the negative of the original product function, but a new function

$$\psi' = \phi_{1s} (2) \, \bar{\phi}_{1s} (1) \, \phi_{2s} (3).$$

3

Only the combination $(\psi - \psi')$ would be antisymmetric for the exchange of electrons (1) and (2). If we now want to make our wave function antisymmetric for the exchange of any two particles, more terms like this have to be added or subtracted until we have six products, the whole thing being such that it may be represented by the shorthand notation of a determinant, called a Slater determinant,

$$\Psi = \frac{1}{\sqrt{6}} \begin{vmatrix} \phi_{1s}(1) & \bar{\phi}_{1s}(1) & \phi_{2s}(1) \\ \phi_{1s}(2) & \bar{\phi}_{1s}(2) & \phi_{2s}(2) \\ \phi_{1s}(3) & \bar{\phi}_{1s}(3) & \phi_{2s}(3) \end{vmatrix}$$

More usually one just writes down the diagonal of the determinant, it being understood that the product has to be antisymmetrized and multiplied by a normalizing factor,

i.e.
$$\Psi = |\phi_{1s} \bar{\phi}_{1s} \phi_{2s}|$$

1.3.   The expansion of orbitals

Our orbital ϕ, which in the present work will most frequently be a molecular orbital, is normally expanded in terms of a set of atomic orbitals,

$$\phi = \sum_i c_i \, \chi_i$$

The set of a.o.s is called the basis set.

Our problem is to determine the coefficients c_i . The χ_i can be chosen to form an orthonormal set,

$$\int \chi_i^2 \, dv = 1,$$

$$\int \chi_i \chi_j \, dv = 0 \quad \text{for} \quad i \neq j.$$

In practice it is convenient enough if they are normalized but not orthogonal, i.e. $\int \chi_i^2 \, dv = 1$, but $\int \chi_i \chi_j \, dv = S_{ij}$.

1.4.   Matrix form of the Schrödinger equation

When the one-electron wave function ϕ is expressed as a linear combination of basis orbitals χ_i , the wave equation may be written

$$H \sum_i c_i \chi_i = E \sum_i c_i \chi_i ,$$

where H is a one-electron Hamiltonian.

4

We can multiply both sides of this equation by X_j and integrate over all electron coordinates, which will give us the well-known set of secular equations :

$$\sum_i c_i \left(\int X_j H X_i \, dv - E \int X_i X_j \, dv \right) = 0.$$

We write H_{ij} for $\int X_i H X_j \, dv = \int X_j H X_i \, dv$ and S_{ij} for $\int X_i X_j \, dv = \int X_j X_i \, dv$. H_{ij} is frequently called a *matrix element*, the term being synonymous with integral. It can be shown that $H_{ij} = H_{ji}$. The proof is given in several standard texts.[1-5] These equations may be written in the compact matrix form

$$\sum_i c_i \left(H_{ji} - ES_{ji} \right) = 0$$

For such a set of equations to have a non-trivial set of solutions the condition

$$\det \left| H_{ji} - ES_{ji} \right| = 0$$

must hold.

Molecular orbital calculations always make use of this simple equation. In the Hückel theory the precise form of H is not investigated and all the integrals represented by H_{ij} and S_{ij} are either put equal to zero or approximated by semi-empirical estimates.

In the more sophisticated *ab initio* methods, the Hamiltonian H is precisely defined as a self-consistent field Hamiltonian and all the matrix elements are computed. Nonetheless, formally the working equations look very similar to each other.

1.5. Simplification of the secular equations

The secular equations may be simplified using the theorem that states that if an operator R commutes with the total Hamiltonian H (i.e. $RH - HR = 0$) and if eigenfunctions of this operator are used to build elements of the energy matrix, then non-zero elements occur only between functions that correspond to the same eigenvalue of R.

In molecular problems the commuting operators R are normally angular momentum operators or symmetry operators. The use of these operators is a fairly complicated business and is fully treated in specialist works.[2-5] The experimental chemist may be unfamiliar with group theory, but he is usually aware of distinctions between electronic states made on the basis of symmetry. The very widely used classification of electronic states in terms of spin, angular momentum, and symmetry properties is the result of applying the commuting operator theorem.

We divide states into types such as $^1\Sigma^+$ for closed-shell diatomics with no resultant spin or angular momentum and a symmetric wave function, or ^3B_u for a particular excited state of benzene.

From the point of view of molecular orbital calculations it is necessary that the wave functions reflect the symmetry and angular momentum properties of the state in question, but in addition the use of the theorem enables us to put many matrix elements equal to zero. If Ψ and H now refer to the totality of electrons,

$$\int \Psi_1^* H \Psi_2 \, d\tau \equiv \langle \Psi_1 | H | \Psi_2 \rangle$$

will be equal to zero if Ψ_1 and Ψ_2 refer to states that are different species, for example one a singlet and the other a triplet or one a Σ state and the other a Π state.

In this way the set of secular equations may be simplified and large matrices reduced to a series of blocks along the diagonal, e.g.

$$\begin{vmatrix} (H_{11} - E) & H_{21} & & & \\ H_{12} & (H_{22} - E) & 0 & & \\ & & (H_{33} - E) & & \\ 0 & & & (H_{44} - E) & H_{45} \\ & & & H_{54} & (H_{55} - E) \end{vmatrix} = 0$$

Each small block is a separate secular equation problem for states of a particular symmetry species and can be treated separately.

When integrals of the form $\langle \Psi_1 | H | \Psi_2 \rangle$ are zero, we say that there is no 'off-diagonal matrix element' between 1 and 2.

The matrix elements referred to here are between wave functions for complete states. These can be expanded into matrix elements between molecular symmetry orbitals. A molecular symmetry orbital, such as a σ or π or a_1 type orbital, is a linear combination of a.o.s. The set of a.o.s in each combination again reflects the symmetry of the molecule. This permits each block in the above matrix to be decomposed into a set of smaller blocks, one for each molecular-orbital symmetry type. Even beyond this stage many of the component integrals over basis set a.o.s within one m.o. type can be equated to zero, owing to the symmetry properties of the basis function, e.g. s, p_x, p_y, p_z, etc.

2
MOLECULAR ORBITALS

In molecular orbital theory, the wave function of the molecule consists of an antisymmetrized product of orbitals; one orbital for each individual electron. This gives a Slater determinant. Furthermore, each of the one-electron orbitals is itself a complicated linear combination of atomic orbitals. We thus have a hierarchy of complication.

The wave function of the molecule Ψ can be broken down as follows:

$$\Psi = \mathcal{Q}\, \psi \qquad ; \mathcal{Q} \text{ is an antisymmetrizing operator,}$$

but

$$\psi = \phi_1\, \phi_2 \ldots \phi_n \quad ; \phi_i \text{ is a one-electron spin orbital,}$$

and the spatial part of

$$\phi_i = \sum_k c_{ik}\, \chi_k \quad ; \chi_k \text{ are a.o.s.}$$

Thus although the use of the variation principle for the molecule will involve us in calculating the integrals $<\Psi|H|\Psi>$ and $<\Psi|\Psi>$, these can be broken down into integrals involving the molecular orbitals, which in turn reduce, or, more honestly, expand, to atomic integrals involving the a.o.s χ_k.

The detailed rules that simplify the calculation of such matrix elements will be treated fully in Chapter 4, but in this chapter some very simple systems will be treated in order to show how a molecular calculation will reduce to a problem of atomic integrals. Further, the examples will serve to introduce some more standard notation.

2.1. Atomic units

There is considerable advantage, even at the level of writing down equations, if one works in atomic units. These take as fundamental quantities

the mass of the electron m_e	as the unit of mass,
the charge of the electron e	as the unit of charge,
the Bohr radius a_0	as the unit of length,
$e^2/4\pi\epsilon_0 a_0$	as the unit of energy ($\equiv 27 \cdot 21$ eV).

7

In these units Planck's constant $h = 2\pi$ and hence $\hbar = 1$ and $8\pi^2 m/h^2 = 2$. Thus for the hydrogen atom the wave equation may be written

$$\left\{ -\frac{1}{2} \nabla^2 - \frac{1}{r} \right\} \psi = E\psi$$

Here ∇^2 is the kinetic energy operator

$$\left(\frac{\partial^2}{\partial x^2} + \frac{\partial^2}{\partial y^2} + \frac{\partial^2}{\partial z^2} \right)$$

2.2. The hydrogen-molecule ion

The starting point for all molecular calculations is the Schrödinger equation

$$H\Psi = E\Psi.$$

In this case if we label the nuclei A and B, so that r_A is the distance of the electron from A and r_B from B, we have for the wave equation

$$\left\{ -\frac{1}{2} \nabla^2 - \frac{1}{r_A} - \frac{1}{r_B} \right\} \psi = E_{el}\psi.$$

The three energy terms in the Hamiltonian are

$$-\frac{1}{2} \nabla^2 \;,\quad \text{the kinetic energy of the electron,}$$

$$-\frac{1}{r_A} \;,\quad \text{the coulomb attraction between the nucleus } A \text{ and the electron,}$$

$$-\frac{1}{r_B} \;,\quad \text{the coulomb attraction between the nucleus } B \text{ and the electron}$$

The energy E_{el} is thus the electronic energy. To this we need to add the nuclear repulsion, which in terms of atomic units is just $\frac{1}{R}$,

$$\text{i.e.} \quad E_{total} = E_{el} + \frac{1}{R},$$

The subscript on E_{el} is normally left out.

The nuclei are invariably assumed to be fixed. We do separate calculations for each configuration of the nuclei. The calculated energy is the electronic part, to which is added the internuclear repulsion energy for the particular geometry chosen. (This separation of energies is a consequence of the Born–Oppenheimer approximation.)

8

2.3. Solution of the wave equation for H_2^+ by the LCAO method

There are several ways of setting about the solution of the wave equation for the hydrogen-molecule ion. For our purposes the method in which the molecular orbitals are chosen to be linear combinations of atomic orbitals is the most interesting. This is usually referred to as the LCAO method.

We represent our molecular orbital ϕ as a sum of the $1s$ atomic orbitals on the two nuclei. Since our problem concerns only a single electron there is no distinction between the wave function for the single electron ϕ and the molecular wave function ψ.

The two atomic orbitals will have equal coefficients by symmetry, so the orbital ϕ is

$$\phi = N(1s_A + 1s_B).$$

A and B refer to the two nuclei and N is a normalizing constant given by

$$N = \frac{1}{\sqrt{\{2(1 + S)\}}}$$

where $\quad S = \int 1s_A\, 1s_B\, dv,$

The energy of the orbital and hence the electronic energy of the system will be given by the formula

$$E = \frac{<\psi|H|\psi>}{<\psi|\psi>}$$

Substituting the electronic Hamiltonian for H_2^+ and the LCAO expression for ψ into this formula, we obtain

$$E = \frac{1}{2(1 + S)} \int (1s_A + 1s_B) \left\{ -\frac{1}{2}\nabla^2 - \frac{1}{r_A} - \frac{1}{r_B} \right\} (1s_A + 1s_B)\, dv$$

We have, by symmetry,

$$\int 1s_A \left(-\frac{1}{2}\nabla^2\right) 1s_A\, dv = \int 1s_B \left(-\frac{1}{2}\nabla^2\right) 1s_B\, dv$$

and

$$\int 1s_A \frac{1}{r_A} 1s_A\, dv = \int 1s_B \frac{1}{r_B} 1s_B\, dv.$$

The full expression for the orbital energy may therefore be simplified to

$$E = \frac{1}{(1 + S)} \left[\left[\int 1s_A \left(-\frac{1}{2} \nabla^2 \right) 1s_A \, dv + \int 1s_A \left(-\frac{1}{2} \nabla^2 \right) 1s_B \, dv - \right. \right.$$

$$\left. - \int 1s_A \frac{1}{r_A} 1s_A \, dv - \int 1s_A \frac{1}{r_B} 1s_A \, dv - 2 \int 1s_A \frac{1}{r_B} 1s_B \, dv \right].$$

For $1s_A$ and $1s_B$ we can use hydrogen-atom solutions

$$1s_A = \frac{1}{\sqrt{\pi}} e^{-r_A} \quad \text{and} \quad 1s_B = \frac{1}{\sqrt{\pi}} e^{-r_B}$$

This leaves us with a set of integrals to calculate. Some are trivial and some rather tedious, but all can be computed by standard methods in milliseconds on even very small computers. When the integrals have been computed one adds them up to find the energy.

2.4. The hydrogen molecule H_2

The hydrogen-molecule ion is very much a special case, since it only contains a single electron. Far more typical is the neutral molecule. The wave equation may be written in atomic units as

$$\left\{ -\frac{1}{2} \nabla_1^2 - \frac{1}{2} \nabla_2^2 - \frac{1}{r_{1A}} - \frac{1}{r_{1B}} - \frac{1}{r_{2A}} - \frac{1}{r_{2B}} + \frac{1}{r_{12}} \right\} \Psi_{(1,2)} = E \Psi_{(1,2)}$$

where subscripts 1 and 2 refer to the two electrons and A and B to the two nuclei.

Again, $E_{\text{total}} = E_{\text{el}} + \frac{1}{R}$, R being the internuclear distance.

The equation may be simplified by noting that apart from the $\frac{1}{r_{12}}$ term the Hamiltonian is a sum of two H_2^+ Hamiltonians,

i.e.
$$\left\{ H_{(1)}^N + H_{(2)}^N + \frac{1}{r_{12}} \right\} \Psi_{(1,2)} = E \Psi_{(1,2)}$$

If the electron repulsion was neglected and H just equalled $H_{(1)}^N + H_{(2)}^N$ we could replace $\Psi_{(1,2)}$ by a product of two one-electron functions ϕ_1 and ϕ_2. These one-electron functions or orbitals would be simply eigenfunctions of the equation

$$H_{(1)}^N \phi_1 = \epsilon_1 \phi_1.$$

where $H_{(1)}^N$ is a hydrogen-molecule ion Hamiltonian.

This idea of building up molecular wave functions as products of one-electron solutions for H_2^+ corresponds exactly to the familiar notion in atoms of saying 'the structure of the beryllium atom is $1s^2\ 2s^2$' — the $1s$ and $2s$ being derived from hydrogen-atom solutions.

For H_2 then we can say that the m.o. configuration is $1\sigma_g^2$ or $1\sigma_g^\alpha\ 1\sigma_g^\beta$ — two electrons with opposite spins in the lowest orbital of the type obtained in solving the wave equation for H_2^+. The orbital is labelled $1\sigma_g$ for symmetry reasons. In the notation introduced in the last chapter we could say the wave function as a Slater determinant is

$$\Psi = |\ 1\sigma_g\ 1\bar{\sigma}_g\ |.$$

This is the shorthand form of

$$\frac{1}{\sqrt{2}} \begin{vmatrix} 1\sigma_g(1) & 1\bar{\sigma}_g(1) \\ 1\sigma_g(2) & 1\bar{\sigma}_g(2) \end{vmatrix} \equiv \frac{1}{\sqrt{2}}\ \left[1\sigma_g(1)\ 1\bar{\sigma}_g(2) - 1\sigma_g(2)\ 1\bar{\sigma}_g(1) \right].$$

The molecular orbital $1\sigma_g$ could be expressed in LCAO form as

$$\frac{1}{\sqrt{2}}\ (1s_A + 1s_B).$$

However we may add even more terms to this expansion of $\phi_{1\sigma_g}$ if we wish,

e.g. $\phi_{1\sigma_g} = c_1\ 1s_A + c_2\ 1s_B + c_3\ 2s_A + c_4\ 2s_B + c_5\ 2p_{xA} + \ldots,$

but whatever the length of the LCAO expression we can always express the energy (or any other expectation value) as a sum of one-or two-electron integrals over molecular orbitals ϕ_i. There are rules — Slater's rules (see Chapter 4) which enable this to be done quite simply even in complicated cases, but for H_2 the grand expression can be written out in full and it is illustrative to do so.

The molecular wave function Ψ is $|\ 1\sigma_g\ 1\bar{\sigma}_g\ |$. Here $1\sigma_g$ is a linear combination of a.o.s, which are mathematical functions in space coordinates, generally polar coordinates r, θ, and ϕ, but here just functions of r, there being no angular variation, so that the function is symmetric about the molecular axis. This orbital becomes a spin orbital by multiplying by one of the two spin functions α or β, which are orthonormal, i.e.

$$\int \alpha^* \alpha\, ds = 1 \quad \int \beta^* \beta\, ds = 1 \quad \int \alpha^* \beta\, ds = \int \beta^* \alpha\, ds = 0$$

Here s is the spin variable, which can take only two values, so that the integral is really a summation.

The energy of the ground state $= \; <\Psi \,|\, H \,|\, \Psi>$

$$= \frac{1}{2} \Big\langle 1\sigma_g\,(1)\;1\bar{\sigma}_g\,(2) \;-$$

$$1\sigma_g\,(2)\;1\bar{\sigma}_g\,(1)\; \Big|\, H^N_{(1)} + H^N_{(2)} + \frac{1}{r_{12}}\, \Big|\; 1\sigma_g\,(1)\;1\bar{\sigma}_g\,(2)\; -\; 1\sigma_g\,(2)\;1\bar{\sigma}_g\,(1)\Big\rangle$$

We will now discuss the various integrals individually.
The first term in the expansion of the above expression will be

$$< 1\sigma_g\,(1)\;1\bar{\sigma}_g\,(2)\; |\, H^N_{(1)} \,|\; 1\sigma_g\,(1)\;1\bar{\sigma}_g\,(2) >$$

If we separate electrons 1 and 2, this may be rewritten as

$$\int 1\sigma_g\,(1)\, H^N_{(1)}\; 1\sigma_g\,(1)\; \mathrm{d}\tau_1,\; \int 1\bar{\sigma}_g\,(2)\; 1\bar{\sigma}_g\,(2)\; \mathrm{d}\tau_2,$$

since H^N_1 operates only on electron 1. Then if we further separate space and spin parts it becomes

$$\underbrace{\int 1\sigma_g\,(1)\, H^N_{(1)}\; 1\sigma_g\,(1)\; \mathrm{d}v_1}_{\epsilon^N_{1\sigma_g}} \times \underbrace{\int \alpha(1)\,\alpha(1)\; \mathrm{d}s_1}_{1}$$

$$\times \underbrace{\int 1\sigma_g\,(2)\; 1\sigma_g\,(2)\; \mathrm{d}v_2}_{1} \times \underbrace{\int \beta(2)\,\beta(2)\,\mathrm{d}s_2}_{1}$$

Thus the whole integral is reduced to the single term $\epsilon^N_{1\sigma_g}$ ($\epsilon^N_{1\sigma_g}$ is is defined as $\int 1\sigma_g\,(1)\, H^N_1\; 1\sigma_g\,(1)\; \mathrm{d}v_1$). Similarly

$$< 1\sigma_g\,(1)\;1\bar{\sigma}_g\,(2)\, |\, H^N_{(2)}\, |\; 1\sigma_g\,(1)\;1\bar{\sigma}_g\,(2)> \;=\; \epsilon^N_{1\sigma_g}.$$

There will be four such terms, so that when the sum of these integrals is multiplied by the factor $\frac{1}{2}$ we are left with $2\epsilon^N_{1\sigma g}$. There then remain the terms involving the operator $\dfrac{1}{r_{12}}$. The first of these is

$$\Big\langle 1\sigma_g\,(1)\;1\bar{\sigma}_g\,(2)\; \Big|\, \frac{1}{r_{12}}\, \Big|\; 1\sigma_g\,(1)\;1\bar{\sigma}_g\,(2)\Big\rangle .$$

Separating as before we get

$$\underbrace{\int\!\!\int 1\sigma_g\,(1)\,1\sigma_g\,(2)\, \frac{1}{r_{12}}\, 1\sigma_g\,(1)\,1\sigma_g\,(2)\,\mathrm{d}v_1\,\mathrm{d}v_2}_{J_{1\sigma_g 1\sigma_g}} \quad \underbrace{\int \alpha(1)\,\alpha(1)\;\mathrm{d}s_1}_{1} \quad \underbrace{\int \beta(2)\,\beta(2)\;\mathrm{d}s_2}_{1}$$

$J_{1\sigma_g\,1\sigma_g}$ is defined in this way, and if we write electron 1 on one side of the operator and electron 2 on the other it is clear that the integral represents the coulomb interaction between electron clouds due to the two electrons separated by a distance of r_{12}. Thus equivalent definitions are

$$J_{1\sigma_g 1\sigma_g} = \int\int 1\sigma_g^2(1) \frac{1}{r_{12}} 1\sigma_g^2(2) \, dv_1 \, dv_2$$

$$= \int\int 1\sigma_g^*(1) 1\sigma_g(1) \frac{1}{r_{12}} 1\sigma_g^*(2) 1\sigma_g(2) \, dv_1 \, dv_2$$

$$= \int\int 1\sigma_g^*(1) 1\sigma_g^*(2) \frac{1}{r_{12}} 1\sigma_g(2) 1\sigma_g(1) \, dv_1 \, dv_2.$$

The cross term will vanish owing to spin orthogonality, e.g.

$$\left\langle 1\sigma_g(1) 1\bar\sigma_g(2) \frac{1}{r_{12}} 1\sigma_g(2) 1\bar\sigma_g(1) \right\rangle =$$

$$= \int\int 1\sigma_g(1) 1\sigma_g(2) \frac{1}{r_{12}} 1\sigma_g(2) 1\sigma_g(1) \, dv_1 \, dv_2$$

$$\times \underbrace{\int \alpha(1) \beta(1) \, ds_1}_{0} \times \underbrace{\int \alpha(2) \beta(2) \, ds_2}_{0}$$

The two coulomb terms when multiplied by the factor ½ give us a single J. The entire expression for the energy reduces to

$$E\left(H_2 : {}^1\Sigma_g^+\right) = 2\,\epsilon_{1\sigma_g}^N + J_{1\sigma_g 1\sigma_g}$$

This type of procedure may be generally followed but where there are more electrons the number of terms is vast and a very large piece of paper is required. The final result is, however, very simple and perfectly understandable in words :

'the energy of H_2 is twice the sum of the energy we would have if there was only one electron in the molecule, plus the coulomb interaction between the two electrons' — a one-electron integral and a two-electron integral.

2.5. The triplet state of H_2

The ground state of H_2 ($X\ {}^1\Sigma_g^+$) has the configuration $1\sigma_g^2$. If we excite one of these electrons to the next unfilled orbital, $1\sigma_u$, and both electrons have the same spin, a configuration $1\sigma_g 1\sigma_u$, the resulting state is a triplet, ${}^3\Sigma_u^+$ being its spectroscopic designation. The wave function $\Psi = |1\sigma_g 1\sigma_u|$, so that the energy will be

13

given by $\langle |1\sigma_g\ 1\sigma_u||H||1\sigma_g\ 1\sigma_u|\rangle$, the denominator being unity. It is instructive to expand this as we did for the ground state since we encounter a further type of integral not found for the ground state. Since both electrons are of the same spin, say α, no terms will disappear owing to spin orthogonality. Thus we have

$$E = \frac{1}{2} \left\langle 1\sigma_g\ (1)\ 1\sigma_u\ (2) - 1\sigma_g\ (2)\ 1\sigma_u\ (1)\ \left| H^N_{(1)} + H^N_{(2)} + \frac{1}{r_{12}} \right| \right.$$

$$\left. 1\sigma_g\ (1)\ 1\sigma_u\ (2) - 1\sigma_g\ (2)\ 1\sigma_u\ (1) \right\rangle .$$

Most of the terms will behave just as the ground-state expansion except for the cross term

$$-\left\langle 1\sigma_g\ (1)\ 1\sigma_u\ (2)\ \left| \frac{1}{r_{12}} \right| 1\sigma_g\ (2)\ 1\sigma_u\ (1) \right\rangle$$

$$-\ -\ \underbrace{\int\int 1\sigma_g(1)\ 1\sigma_u\ (2)\frac{1}{r_{12}}\ 1\sigma_g\ (2)\ 1\sigma_u(1)\ dv_1\ dv_2}_{K_{1\sigma_g\ 1\sigma_u}}$$

$$\times\ \underbrace{\int \alpha(1)\ \alpha(1)\ ds_1}_{1}\ \times \underbrace{\int \alpha(2)\ \alpha(2)\ ds_2}_{1}$$

Alternatively $K_{1\sigma_g\ 1\sigma_u}$ may be written as

$$\int\int 1\sigma_g\ (1)\ 1\sigma_u\ (1)\ \frac{1}{r_{12}}\ 1\sigma_g\ (2)\ 1\sigma_u\ (2)\ dv_1\ dv_2\ ,$$

the important thing to notice being that in a K (exchange) integral electron 1 is in two different orbitals as is electron 2, so that the interaction is not just a simple electrostatic repulsion.

The complete energy expression for the $^3\Sigma_u^+$ state is then

$$E(H_2 : {}^3\Sigma_u^+) = \epsilon^N_{1\sigma_g} + \epsilon^N_{1\sigma_u} + J_{1\sigma_g\ 1\sigma_u} - K_{1\sigma_g\ 1\sigma_u}$$

The two energy expressions obtained for the two states of H_2 illustrate the quite general form of energy expressions for molecules. The total energy is the sum of the one-electron energies (the energy each electron would have were it the only electron in the molecule) plus a coulomb integral (the interaction between the charge clouds) for every pair of electrons, minus an exchange integral for every pair of electrons in the molecule which have the same spin. The first two types of term can be immediately understood from purely classical electrostatic ideas, but exchange is a purely quantum mechanical effect with no physical meaning, resulting from the Pauli Principle, which ensures that the wave function is antisymmetric. It is the integrals between different product terms in the sum of products which give rise to K integrals as is shown in the above case.

2.6. Expansion of the molecular integrals

We have seen how the energy of the H_2 ground state is a sum of the two molecular integrals $\epsilon^N_{1\sigma_g}$ and $J_{1\sigma_g 1\sigma_g}$. These may be expanded in terms of atomic integrals.

If
$$1\sigma_g = \frac{1}{\sqrt{\{2(1 + S)\}}}(1s_A + 1s_B),$$

then

$$\epsilon^N_{1\sigma_g} = \frac{1}{2(1 + S)} \int (1s_A + 1s_B)\left[-\frac{1}{2}\nabla^2 - \frac{1}{r_A} - \frac{1}{r_B}\right](1s_A + 1s_B)\,dv;$$

similarly $J_{1\sigma_g 1\sigma_g}$ reduces (or rather increases) to

$$\frac{1}{4(1 + S)^2} \int (1s_A + 1s_B)^2_{(1)}\frac{1}{r_{12}}(1s_A + 1s_B)^2_{(2)}\,dv_1\,dv_2$$

Clearly if $1\sigma_g$ was expanded as a longer sum of atomic integrals and if there were many more electrons, the number of atomic integrals required would grow very rapidly. Indeed for quite small molecules the number of atomic integrals required may be as many as several million. This is why it was not until computers that were fast and had large stores became available that *ab initio* calculations made very much impact. However the basic theory is simple and its execution becomes routine. In their simplest form the programmes are not very difficult to write, although this is not often necessary, as specialists have made their programmes available. These computer programmes involve the very efficient use of standard techniques and careful shunting of numbers from one part of the computer to another. However, the programmes are now so sophisticated that many man—years are required for their development.

2.7. General LCAO method for diatomic molecules

In H_2^+ we constructed the $1\sigma_g$ m.o. by combining the two 1s hydrogen a.o.s. Pictorially,

$$1\sigma_g = 1s_A + 1s_B \equiv$$

and

$$1\sigma_u = 1s_A - 1s_B \equiv$$

15

For a general homonuclear molecule we can represent the m.o.s by the following well-known type of diagram.

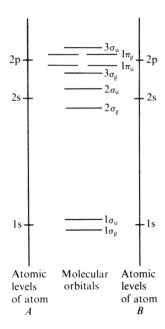

Atomic levels of atom A — Molecular orbitals — Atomic levels of atom B

The m.o.s are labelled σ or π depending on whether they are symmetrical about the internuclear axis or have a nodal plane passing through the nuclei and g or u depending on whether they remain unchanged or simply change sign when inverted at the centre of symmetry. This labelling is familiar, but the numbering $1\sigma_g$, $2\sigma_g$, $3\sigma_g$, etc. is a modern convention, replacing older versions which tended to indicate some atomic parentage of the m.o. This has been abandoned and the orbitals of a particular type are just numbered consecutively from the bottom of the diagram.

For heteronuclear molecules there is no longer a centre of symmetry, so there is no g or u, but there are still σ and π and the running number, i.e. (See figure opposite).

The reason for the new notation is that we may express for example 1σ as a very long linear combination of a.o.s of the correct symmetry, e.g. for CO

$$1\sigma = c_1\, 1s_c + c_2\, 1s_0 + c_3\, 2s_c + c_4\, 2p_{\sigma_c} + \dots$$

The more terms the better the energy, by the variation principle, but the less clear the origin of the major constituents.

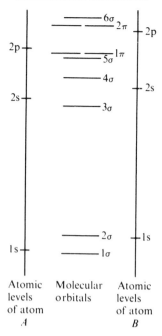

One further point to notice is the double degeneracy of the π orbital formed in the simplest case (H_2^+) by $(2p_{x_A} + 2p_{x_B})$ and $(2p_{y_A} + 2P_{y_B})$, the z axis being taken as the internuclear axis. We can then have π_x and π_y degenerate orbitals. It is, however, more common and convenient to work with the complex forms $\pi^+(=\pi_x + i\pi_y)$ and $\pi^-(=\pi_x - i\pi_y)$, which contain factors $e^{+i\phi}$ and $e^{-i\phi}$ respectively.

For two of the molecules we are going to take as examples the configurations of the ground states are

$$N_2 : 1\sigma_g^2\ 1\sigma_u^2\ 2\sigma_g^2\ 2\sigma_u^2\ 3\sigma_g^2\ 1\pi_u^4$$

$$CO : 1\sigma^2\ 2\sigma^2\ 3\sigma^2\ 4\sigma^2\ 5\sigma^2\ 1\pi^4$$

The wave functions will be 14×14 determinants, each term of which is a linear combination of atomic orbitals.

To formulate the energy expression for a molecule the procedure is as follows

(a)   Write down the Hamiltonian — in general this is

$$H = \left\{ \sum_i \left(-\frac{1}{2}\nabla_i^2\right) - \sum_{i,\mu} \frac{Z_\mu}{r_{i\mu}} + \sum_{i<j} \frac{1}{r_{ij}} \right\}$$

where i and j are electron indices and μ nuclei.

(b) Express the wave function Ψ as a product, which has to be antisymmetrized.

(c) Express each ϕ_i as an LCAO,

$$\phi_i = \sum_k c_{ik} \chi_k$$

(d) Find the energy (or other mean value) in terms of integrals over molecular orbitals (ϕ_i).

(e) Expand the integrals over m.o.s into integrals over a.o.s (χ).

3

SELF-CONSISTENT FIELDS

We have seen that, if each m.o. is a linear combination

$$\phi_i = \sum_k c_{ik} \chi_k,$$

we can ultimately express all we need to calculate in terms of integrals involving the χ_k, multiplied, of course, by the appropriate sums and products of the LCAO coefficients c_{ik}. Indeed the only snag in the scheme outlined at the end of the previous chapter is, how do we know what the values of c are?

The answer to this is that we use the idea of self-consistent field orbitals.

3.1. The Hartree equations

The Hartree self-consistent field equations are based on the fact that if the wave function for the molecule is just a single product of orbitals, then the energy is the sum of the one-electron energies (kinetic energy and electron nuclear attractions) and coulomb interactions between the charge clouds of all pairs of electrons i and j.

$$E = \sum_i \epsilon_i^N + \sum_{i<j} \phi_i^2 (1) \frac{1}{r_{12}} \phi_j^2 (2) \, dv_1 \, dv_2 .$$

The condition that this energy should be a minimum, an application of the variation theorem, together with the auxiliary conditions

$$\int \phi_i \phi_j \, dv_i \, dv_j = 1 \quad \text{if} \quad i = j \quad \text{or} \quad 0 \quad \text{if} \quad i \neq j$$

gives us the Hartree equations for the 'best orbitals',

$$\left\{ H^N + \sum_{j=1}^n \left(\int \phi_j^2 (2) \frac{1}{r_{12}} dv_2 \right) \right\} \phi_i (1) = \sum_j \epsilon_{ij} \phi_j (1),$$

each ϕ_i being a linear combination of functions χ.

The Hamiltonian contains terms involving the ϕ_i, which we wish to calculate. We thus have to use an iterative method of solution.

The Hartree equations are not now normally used since they are

19

based on the idea of the wave function for the molecule being a single product of one-electron orbitals and not an antisymmetrized product.

3.2. The Hartree–Fock equations

These do consider the wave function to be an antisymmetrized product and enable one to reach the best Slater-determinant solution iteratively.

As mentioned in the last chapter, if we include antisymmetry, then the energy is a sum of one-electron, coulomb, and exchange terms

$$E = \sum_i \epsilon_i^N + \sum_{i<j} \int \phi_i^2 (1) \frac{1}{r_{12}} \phi_j^2 (2) \, dv_1 \, dv_2 - $$

same spin (o.⊂. ø)

$$- \sum_{i<j}' \int \phi_i (1) \phi_j (1) \frac{1}{r_{12}} \phi_i (2) \phi_j (2) \, dv_1 \, dv_2$$

– all orbitals assumed real.

Or for closed-shell molecules, where all the electrons are paired with others of opposite spin,

$$E = \sum_i 2 \epsilon_i^N + \sum_{i,j} (2J_{ij} - K_{ij}),$$

where here i and j label the orbitals; for example in the diatomic molecule BH, with $\Psi = |1\sigma^2 \, 2\sigma^2 \, 3\sigma^2|$,

$$E = 2\epsilon_{1\sigma}^N + 2\epsilon_{2\sigma}^N + 2\epsilon_{3\sigma}^N + J_{1\sigma1\sigma} + 4J_{1\sigma2\sigma} + 4J_{1\sigma3\sigma} + J_{2\sigma2\sigma} + 4J_{2\sigma3\sigma} +$$

$$+ J_{3\sigma3\sigma} - 2K_{1\sigma2\sigma} - 2K_{1\sigma3\sigma} - 2K_{2\sigma3\sigma}$$

(N.B. from the definition on p. 14 when $i=j$, $K_{ii} = J_{ii}$).

The Hartree–Fock equations are obtained by finding the condition for the energy to be a minimum, $\delta E = 0$, and at the same time demanding that the molecular orbitals obtained shall be orthonormal,

$$\langle \phi_i | \phi_j \rangle = \delta_{ij} .$$

This use of the variation principle is strictly applicable only to the lowest state of a given symmetry.

The resulting equations are (for closed shells and real orbitals)

$$\left\{ H^N + \sum_{j=1}^n \int \phi_j^2 (1) \frac{1}{r_{12}} \, dv_2 \right\} \phi_i (1)$$

$$- \left\{ \sum_{j=1}^n{}' \int \phi_j (2) \phi_i (2) \frac{1}{r_{12}} \, dv_2 \right\} \phi_j (1) = \epsilon_i^{SCF} \phi_i (1)$$

or, in the shorthand form,

$$\left\{ H^N + \sum_j J_j - \sum_j' K_j \right\} \phi_i(1) = \epsilon_i^{SCF} \phi_i(1),$$

Here we have used coulomb and exchange operators which are defined as

$$J_j \, \phi_i(1) = \left(\int \phi_j^2(2) \frac{1}{r_{12}} \, dv_2 \right) \phi_i(1)$$

$$K_j \, \phi_i(1) = \left(\int \phi_j(2) \, \phi_i(2) \frac{1}{r_{12}} \, dv_2 \right) \phi_j(1).$$

It should be noted that the J and K operators as opposed to J and K integrals have a single subscript rather than two.

In an even more compact form we may write

$$H^{SCF} \, \phi_i(1) = \epsilon_i^{SCF} \, \phi_i(1).$$

Again since the Hamiltonian contains the answer we are seeking, the set of equations, one for each ϕ_i, has to be solved iteratively.

For atoms where the same equations apply but we have the simplifying property of spherical symmetry, the Hartree–Fock equations can be solved numerically, giving an exact solution to the equations — not of course an exact solution to the problem, as there are assumptions inherent in the derivation of the equations which will be discussed later (Chapter 11). The exact solution of the equations is equivalent to taking our LCAO expansion to an infinite number of terms, and such a solution is called the Hartree–Fock limit.

Although some progress has been made towards solving molecular Hartree–Fock equations numerically, in general ϕ_i is represented as a linear combination and the equations are solved analytically. The longer the expansion used (the bigger the basis set) the closer the result will come to the Hartree–Fock limit.

3.3. The Roothaan equations

The Hartree–Fock equations in the LCAO approximation are normally called the Roothaan Equations. A non-rigorous but readily comprehensible derivation is as follows.

We have, as above,

$$H^{SCF} \, \phi_i = \epsilon_i^{SCF} \, \phi_i.$$

Now if

$$\phi_i = \sum_n c_{in} \, \chi_n$$

then

$$H^{SCF} \sum_n c_{in} \, \chi_n = \epsilon_i^{SCF} \sum_n c_{in} \, \chi_n$$

and in the familiar manner we can multiply both sides of this equation by say χ_m and integrate over all space.

21

$$\sum_n c_{in} \int \chi_m \, H^{SCF} \, \chi_n \, dv = \epsilon_i^{SCF} \sum_n c_{in} \int \chi_m \, \chi_n \, dv,$$

i.e.

$$\sum_n c_{in} \, (H_{mn}^{SCF} - \epsilon_i^{SCF} \, S_{mn}) = 0.$$

Such a set of equations is only soluble if

$$\det |H_{mn}^{SCF} - \epsilon_i^{SCF} \, S_{mn}| = 0.$$

This secular determinant looks very like that for the simple Hückel m.o. method where the matrix elements H_{mn} and S_{mn} are set equal to empirical parameters or zero.

If H_{mn}^{SCF} and S_{mn} could be calculated the secular determinant could be solved directly for the eigenvalues, the SCF orbital energies ϵ_i^{SCF}. However, both H_{mn} and S_{mn} demand a knowledge of the wave functions we are trying to find and yet again the solution has to be iterative, which makes the use of a computer mandatory.

3.4. An example – LiH

Let us take the simple example of LiH and outline the procedure.
The atoms have the structures Li : $1s^2\,2s$,

$$\text{H} : 1s.$$

The m.o. diagram is then

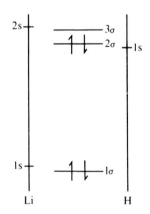

i.e. $$\Psi\left(\text{LiH}; X\,{}^1\Sigma^+\right) = |1\sigma^2\,2\sigma^2|.$$

Now $$\phi_{1\sigma} = c_1\,1s_{Li} + c_2\,1s_H + c_3\,2s_{Li},$$

where, for example, $1s = \dfrac{1}{\sqrt{\pi}}\,e^{-ar}$ – an atomic orbital

In outline our procedure might be:
1.   Guess some values of the c's.

2. Calculate all the various atomic integrals and hence build up $H_{i\sigma j\sigma}^{SCF}$ and $S_{i\sigma j\sigma}$

3. Solve the determinantal equation giving the possible values of ϵ_i^{SCF}

4. Substitute these in the secular equations giving new c's.

5. Go back to stage 1 and repeat until the values of ϵ_i^{SCF} or the c's converge to steady values within an arbitrary threshold ($\sim 10^{-6}$) and then take the values of the converged c's.

The result of this will be to provide SCF orbitals, both occupied and virtual (unoccupied): that is to say the coefficients in

$$\phi_{i\sigma} = c_{i_1} \, 1s_{Li} + c_{i_2} \, 1s_H + c_{i_3} \, 2s_{Li} \quad \text{and} \quad \epsilon_{i\sigma}^{SCF} .$$

This is all very straightforward, but lengthy, just the sort of problem at which computers are so powerful.

4

CALCULATION OF MATRIX ELEMENTS

In order to use the molecular orbitals found from a self-consistent field computation one must be able to calculate matrix elements such as the general energy expression

$$\frac{<\Psi\,|\,H\,|\,\Psi>}{<\Psi\,|\,\Psi>}\;.$$

Here the denominator is unity if the wave function is normalized, and the capital letter Ψ represents a Slater determinant.

In Chapter 2 the evaluation of this expression for H_2 showed how many of the terms disappear due to spin or spatial orthogonality, and the result had a form with a clear rationality about it. To extend such expansions to cases with many more electrons would be fearsome were it not for the fact that these regularities persist and are summarized in a set of rules which will be given and illustrated in this chapter. Before doing this, however, a few more words must be said about the integrals J and K.

4.1. Complex wave functions

For convenience in what has been said so far the molecular orbitals have been assumed to be real and the coulomb integral J_{ij} defined as

$$J_{ij} \;=\; \int\int \phi_i^2(1)\,\frac{1}{r_{12}}\,\phi_j^2(2)\;dv_1\;dv_2\;.$$

This is the simplest way of seeing that it represents the coulomb interaction between the charge densities of electrons (1) and (2).

If the orbitals can be complex this is more properly written

$$\int\int \phi_i^*(1)\;\phi_i(1)\,\frac{1}{r_{12}}\,\phi_j^*(2)\;\phi_j(2)\;dv_1\;dv_2\;,$$

which is the same thing as

$$\int\int \phi_i^*(1)\;\phi_j^*(2)\,\frac{1}{r_{12}}\,\phi_j(2)\;\phi_i(1)\;dv_{12}\;;$$

24

electron (1) is still in orbital ϕ_i and (2) is in ϕ_j. Another common shorthand is

$$\left\langle \phi_i\,\phi_j \left| \frac{1}{r_{12}} \right| \phi_i\,\phi_j \right\rangle,$$

where we understand that the complex parts are written on the left-hand side and the real parts on the right.

Yet other shorthand forms are

$$(ii\,|\,jj) \quad \text{or} \quad \zeta^{\phi_i\,\phi_i}_{\phi_j\,\phi_j},$$

where we keep electron (1) on one side and electron (2) on the other side of the operator which is not written in, and bear in mind that $\phi_i^2\,(1)$ is really $\phi_i^*\,(1)\,\phi_i\,(1)$ if this distinction is necessary.

This wealth of ways of expressing the same thing can cause confusion to those who are unfamiliar with the notation, so we now give all the common equivalent forms of writing $J_{2\sigma1\sigma}$ and $K_{2\sigma1\sigma}$.

$$J_{2\sigma1\sigma} = J_{1\sigma2\sigma} = \int\int \phi_{2\sigma}^*\,(1)\,\phi_{2\sigma}\,(1)\,\frac{1}{r_{12}}\,\phi_{1\sigma}^*\,(2)\,\phi_{1\sigma}\,(2)\,dv_1\,dv_2$$

$$= \langle \phi_{2\sigma}\phi_{1\sigma}|1/r_{12}|\phi_{2\sigma}\phi_{1\sigma}\rangle$$

$$= (2\sigma\,2\sigma\,|\,1\sigma\,1\sigma)$$

$$= \zeta^{2\sigma\,2\sigma}_{1\sigma\,1\sigma}$$

and

$$K_{2\sigma\,1\sigma} = K_{1\sigma2\sigma} = \int\int \phi_{2\sigma}^*(1)\,\phi_{1\sigma}^*\,(1)\,\frac{1}{r_{12}}\,\phi_{2\sigma}\,(2)\,\phi_{1\sigma}(2)\,dv_{12}$$

$$= (2\sigma\,1\sigma\,|\,2\sigma\,1\sigma)$$

$$= \zeta^{2\sigma1\sigma}_{2\sigma1\sigma}.$$

In these examples $\phi_{2\sigma} = \phi_{2\sigma}^*$ as there are no complex parts, so much of the difficulty is purely formal.

Care is, however, necessary when we do have complex parts. An m.o. $1\pi^+$ has an azimuthal portion $e^{+i\phi}$ and $1\pi^-$ has the factor $e^{-i\phi}$. This slight complication of course only applies to diatomic and linear polyatomic molecules where there is axial symmetry.

Here we can distinguish two sorts of integrals involving π electrons, usually labelled 2 or 0, i.e. containing the factors

$$e^{i\phi} \times e^{i\phi} = e^{2i\phi} \quad \text{or} \quad e^{i\phi} \times e^{-i\phi} = e^{0\phi}.$$

Sometimes these cases are called $+$ and $-$.

Thus $\quad J^0_{1\pi 1\pi} = \int\int 1\pi^{+*}(1)\ 1\pi^+(1)\dfrac{1}{r_{12}}\ 1\pi^{+*}(2)\ 1\pi^+(2)\ dv_{12}$,

since $\quad 1\pi^{+*} = 1\pi^-$,

i.e. $\quad J^0_{1\pi 1\pi} = \zeta^{1\pi^+ 1\pi^+}_{1\pi^+ 1\pi^+} = (1\pi^+\ 1\pi^+|\ 1\pi^+\ 1\pi^+)$.

If we use the latter shorthand we can see that the integral will be a J integral if the two spatial parts are identical for electron (1) and for electron (2).

It will be a J^0 if the two exponential arguments are the same and J^2 if they are different.

Likewise we can have K^0 and K^2 integrals

$$K^0_{2\pi 1\pi} = \zeta^{2\pi^+ 1\pi^+}_{2\pi^+ 1\pi^+}$$

$$K^2_{2\pi 1\pi} = \zeta^{2\pi^+ 1\pi^-}_{2\pi^+ 1\pi^-} \ .$$

4.2. Rules for taking matrix elements

If we have a pair of wave functions written as Slater determinants

$$\Psi_1 = |\ 1\sigma\ 1\bar{\sigma}\ 2\sigma\ 1\pi^+|\quad \text{and}\quad \Psi_2 = |\ 1\sigma\ 1\bar{\sigma}\ 1\pi^+ 2\sigma|\ ,$$

how, if at all, do these differ? To answer this we must remember that the shorthand represents a determinant, and as we know from elementary mathematics, if we interchange two columns of a determinant we change the sign.

Thus $\quad \Psi_1 = -\Psi_2$.

It is convenient then when taking matrix elements between determinants firstly to bring the two determinants concerned into the maximum coincidence, remembering that each change of columns involves multiplying the wave function by (-1).

In Appendix 1 there are numerous examples of this process and the application of the following rules, which cover all the possibilities that exist when taking matrix elements between two determinantal wave functions Ψ_A and Ψ_B .

(i) $\quad \Psi_A$ and Ψ_B are identical

The spin orbitals will then be the same in both determinants and the matrix element

$$\langle \Psi_A|H|\Psi_B\rangle \quad \text{will equal}\quad \int|\phi_1^2\ \phi_2^2\ \cdots\ \phi_n^2|H|\phi_1^2\phi_2^2\cdots\phi_n^2|\ d\tau$$

$$= \sum_{k=1}^{n} 2\epsilon_k^N + \sum_{kl}^{nn} J_{kl} - {\sum_{kl}^{nn}}' K_{kl} ,$$

the primed sum indicating summation only over pairs of electrons of the same spin. ~ essentially k,l different

For non-paired electrons :

$$\sum_k^{2n} \bar{\epsilon}_k^N + \sum_{k<l}^{nn} J_{kl}$$

26

$$+ \sum_{n<l<l}^{nn} -K_{kl} \quad \text{same spin}$$

Examples. (a) BH

The wave function for the ground state of BH will be

$$\Psi = |1\sigma^2 2\sigma^2 3\sigma^2|.$$

Therefore using the rule,

$$E = \sum_{i=1}^{3} 2\epsilon_{i\sigma}^{N} + \sum_{i,j=1}^{3} (2J - K)_{i\sigma j\sigma}$$

For practice it is probably worthwhile expanding this by writing down every single term, e.g. $J_{1\sigma 2\sigma}$, etc., in order to verify that the above shortened form does include the correct number of integrals. In the expression we have let both indices i and j range from 1 to 3 so that both $J_{1\sigma 2\sigma}$ and $J_{2\sigma 1\sigma}$ (which are identical) are included.

Further it should be remembered that $K_{i\sigma i\sigma} = J_{i\sigma i\sigma}$.

(b) CO

The energy of the ground state of CO, with

$$\Psi = |1\sigma^2 2\sigma^2 3\sigma^2 4\sigma^2 5\sigma^2 1\pi^4|,$$

$$E(^1\Sigma^+) = \sum_{i=1}^{5} 2\epsilon_{i\sigma}^{N} + 4\epsilon_{1\pi}^{N} + \sum_{i,j=1}^{5} (2J - K)_{i\sigma j\sigma} +$$

$$+ \sum_{i=1}^{5} (8J - 4K)_{i\sigma 1\pi} + 6J_{1\pi 1\pi}^{0} - 2K_{1\pi 1\pi}^{2}.$$

The last two terms could cause confusion so let us write out in full the interactions between the π electrons

$$1\pi^+ \overline{1\pi^+} \ 1\pi^- \overline{1\pi^-} ;$$

there will be six coulomb interactions:

$$1\pi^+ \text{ with } \overline{1\pi^+}$$
$$1\pi^+ \text{ with } 1\pi^-$$
$$1\pi^+ \text{ with } \overline{1\pi^-}$$
$$\overline{1\pi^+} \text{ with } 1\pi^-$$
$$\overline{1\pi^+} \text{ with } \overline{1\pi^-}$$
$$1\pi^- \text{ with } \overline{1\pi^-}$$

or diagrammatically

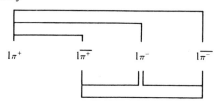

Each bracket represents an interaction.

For the K terms there are only two possible exchange interactions between electrons of the same spin,

$$1\pi^+ \text{ with } 1\pi^-, \text{ giving } \zeta_{1\pi^+ 1\pi^-}^{1\pi^+ 1\pi^-} = K_{1\pi \, 1\pi}^2$$

$$\text{and } \overline{1\pi^+} \text{ with } \overline{1\pi^-}, \text{ giving } \zeta_{\overline{1\pi^+} \, \overline{1\pi^-}}^{\overline{1\pi^+} \, \overline{1\pi^-}} = K_{1\pi \, 1\pi}^2 .$$

This type of matrix element will always be necessary when calculating the energy of a particular state.

(ii) Ψ_A and Ψ_B *have one spin orbital different*

Our matrix element will be the integral

$$\int | \, \dots \, \phi_k^a \, \dots \, |H| \, \dots \, \phi_l^a \, \dots \, | \, d\tau$$

$$= \epsilon_{kl}^N + \sum_m \zeta_{mm}^{kl} - \sum_m {}' \zeta_{lm}^{km} .$$

Here $\epsilon_{kl}^N = \int \phi_k \, H^N \, \phi_l \, dv$. With these integrals particular care must be exercised in deciding which integrals are zero by reasons of spin orthogonality.

Example.   Again using CO as an example, suppose we excite a $5\bar{\sigma}$ electron to a $6\bar{\sigma}$ spin orbital and take the matrix element between this function and the ground state, i.e.

$$\int | \, 1\sigma^2 2\sigma^2 3\sigma^2 4\sigma^2 5\sigma \, 5\bar{\sigma} \, 1\pi^4 |H| \, 1\sigma^2 2\sigma^2 3\sigma^2 4\sigma^2 5\sigma \, 6\bar{\sigma} \, 1\pi^4 | \, d\tau$$

$$= \epsilon_{5\sigma 6\sigma}^N + 2\sum_{i=1}^{4} \zeta_{i\sigma i\sigma}^{5\bar{\sigma} 6\bar{\sigma}} - \sum_{i=1}^{4} \zeta_{6\bar{\sigma} i\bar{\sigma}}^{5\bar{\sigma} i\bar{\sigma}} + 4\zeta_{5\bar{\sigma} 6\bar{\sigma}}^{1\pi \, 1\pi} - 2\zeta_{1\pi 6\bar{\sigma}}^{1\pi 6\pi} + \zeta_{5\bar{\sigma} 6\bar{\sigma}}^{5\sigma 5\sigma} .$$

Such an integral will be needed when doing configuration-interaction calculations and will be returned to in the next chapter. _

(iii) Ψ_A and Ψ_B *differ by two spin orbitals*

In general our integral will be of the type

$$\int | \, \dots \, \phi_k^a \, \dots \, \phi_p^a \, \dots \, |H| \, \dots \, \phi_l^a \, \dots \, \phi_q^a \, \dots \, | \, d\tau = \zeta_{pq}^{kl} - \zeta_{pl}^{kq} .$$

We will only have both of these terms if all the four spin orbitals are of the same spin. Otherwise one or other integral will be zero for reasons of spin orthogonality.

Example.   Again for CO, if both 5σ electrons are excited to the 6σ orbital, taking the matrix element with the ground state ·

$$\int | \, 1\sigma^2 \, \dots \, 4\sigma^2 5\sigma^2 \, 1\pi^4 \, |H| \, 1\sigma^2 \, \dots \, 4\sigma^2 6\sigma^2 1\pi^4 | \, d\tau = \zeta_{5\bar{\sigma} 6\bar{\sigma}}^{5\sigma 6\sigma} - \zeta_{5\bar{\sigma} 6\bar{\sigma}}^{5\sigma 6\bar{\sigma}}$$

the second term

$$\int \phi^{\alpha}_{5\sigma}(1)\, \phi^{\beta}_{6\sigma}(1)\, \frac{1}{r_{12}}\, \phi^{\beta}_{5\sigma}(2)\, \phi^{\alpha}_{6\sigma}(2)\, d\tau \;=$$

$$= \int \phi_{5\sigma}(1)\, \phi_{6\sigma}(1) \frac{1}{r_{12}}\, \phi_{5\sigma}(2)\, \phi_{6\sigma}(2)\, dv \int \alpha(1)\, \beta(1)\, ds_1 \int \beta(2)\, \alpha(2)\, ds_2$$

$$= 0\,.$$

This type of integral is also of great importance in configuration interaction.

(iv) Ψ_A *and* Ψ_B *differ by more than two spin orbitals*

Matrix elements between the two wave functions are then zero. This is because the Hamiltonian only contains terms involving at most two electrons. The proofs for this and the other rules are given in many text-books. In particular one may cite Parr[4], where clear, simple proofs are given. Several examples of the use of the rules are given in Appendix 1. At first one has to be very careful, but with a little practice the use of the rules for taking matrix elements becomes second nature. It usually takes graduate students about one week to master these rules.

5

CLOSED-SHELL CALCULATIONS

We now have enough information to consider performing a closed-shell computation and utilizing the results obtained.

5.1. Basis set of atomic orbitals

We are going to represent each molecular orbital ϕ_i as a linear combination

$$\phi_i = \sum_k c_{ik} \chi_k .$$

The first problem is to decide how many atomic orbitals to use and what algebraic form they have.

Most frequently for diatomic molecules Slater type a.o's are preferred.

$$\chi = r^{n-1} e^{-\zeta r} \times \text{angular part in } \theta \text{ and } \phi.$$

If we take one exponential for each orbital that is filled in the constituent atoms we have what is called a minimal basis set. However, there is in principle no limit to the number of terms we take providing that they are of the correct symmetry. Often two a.o's with slightly different exponents are taken for each filled a.o., such a basis set being referred to as a 'double zeta' basis set. If on the other hand we can take enough orbitals to reproduce the answer obtainable by numerical solution we are at the Hartree—Fock limit and, in general, by gradually augmenting the basis set, the energies converge on this result.

Ideally the orbital exponent ζ should be taken as a variational parameter and the energy of the molecule computed for a series of values until the minimum is found. However, this multivariational problem is extremely time-consuming as each molecular calculation may take several minutes.

Instead the best orbital exponents for the separated atoms are found since atomic calculations can be performed in a fraction of the time required for molecules.

Tables of atomic exponents accurate enough to reproduce the atomic Hartree—Fock energies to six figures in a.u. have been published by Clementi. It is often desirable to add to these functions, which are atomic s and p types, a few d-symmetry orbitals, as these are thought to enable the charge to be concentrated along the bond. This should not be taken as meaning that actual atomic d orbitals are necessarily employed in bonding; rather they are simply convenient variational functions.

The following table gives typical orbital exponents which have been used to compute molecular orbitals and expectation values for N_2 and CO .

(R.K. NESBET, *J. chem. Phys.* **40**, 3619 (1964).)

Type	C	N	O
$1s\sigma$	5·30360	6.21292	7·16512
$1s\sigma$	8·38300	9·36827	10·61430
$2s\sigma$	1·26960	1·46786	1·60111
$2s\sigma$	1·85619	2·24264	2·58881
$2p\sigma$	1·28709	1·52853	1·65145
$2p\sigma$	2·85367	3·33678	3·67544
$3d\sigma$	1·895	1·935	2·103
$2p\pi$	1·28709	1·52853	1·65145
$2p\pi$	2·83567	3·33678	3·67544
$3d\pi$	1·175	1·429	3·019
Atomic SCF energy	− 37·68634	− 54·39754	− 99·39870
Atomic Hartree—Fock limit	− 37·68858	− 54·40091	− 99·40921

5.2. The starting coefficients

Once the size and nature of the basis set has been decided upon, probably by a compromise between the accuracy of computation required and the expense of computer time, the next stage is to guess some initial values of the coefficients c_{ik}. Most *ab initio* programmes contain an efficient extrapolation procedure which uses the result of three successive iterations to obtain a better approximation to the final wave function, so that starting values can be very bad guesses and yet convergence is achieved in about ten cycles. Crude ideas about the major atomic constituents of a particular m.o. may be used. For instance, in a computation of CO the 1σ m.o. is probably going to be largely an a.o. centred on the oxygen, so that this a.o. could have a guessed starting coefficient of 0·9.

Some programmes have as starting data not guesses of the coefficients c_{ik} but estimated sums of products of these,

$$\sum_k c_{ik} c_{jk} \quad .$$

31

The matrix of these is the density matrix for the molecule. This form of starting values of coefficients is convenient since it is in this form that the coefficients are used in computing expectation values, e.g. $J_{1\sigma 2\sigma}$ might be

$$\int\int (c_{11}1s_0 + c_{14} 1s_c + c_{13} + \ldots)^2 \frac{1}{r_{12}} (c_{21} 1s_0 + c_{22} 1s_c + c_{23} \ldots)^2 \, d\tau.$$

However, because of the extrapolation procedure, even the most ludicrous guess of elements of a density matrix will generally not greatly hinder convergence.

5.3. An example of a closed-shell calculation

Having chosen an atomic orbital basis set, guessed coefficients, and used an *ab initio* SCF programme, the resulting output will consist of the molecular-orbital coefficients (i.e. the final coefficients), their orbital energies ϵ_i^{SCF}, and possibly the molecular integrals.

The total electronic energy of the molecule is a sum of SCF orbital energies and one-electron energies for a closed-shell molecule

This may be illustrated by taking $CO\,(X\,^1\Sigma^+)$ again as an example.

$$\Psi = |\,1\sigma^2\,2\sigma^2\,3\sigma^2\,4\sigma^2\,5\sigma^2\,1\pi^4\,|$$

Energy $= \langle\Psi|H|\Psi\rangle$, where H is the electronic Hamiltonian appropriate to the problem, the sum of kinetic and electrostatic terms.

As shown in Chapter 4 this integral is equal to

$$\sum_{i=1}^{5}\epsilon_{i\sigma}^{N} + 4\epsilon_{1\pi}^{N} + \sum_{ij=1}^{5} (2J - K)_{i\sigma j\sigma} + \sum_{i=1}^{5} (8J - 4K)_{1\pi i\sigma} + 6J_{1\pi 1\pi}^{0} - 2K_{1\pi 1\pi}^{2}$$

Now the SCF equation which has been solved is

$$\left\{ H^N + \sum J_i - \sum{}' K_i \right\} \phi_i\,(1) = \epsilon_i^{SCF}\phi_i \tag{1}$$

This expression has been discussed above and is to be found in most books on quantum chemistry. It is, however, rather stark and possibly more readily understood by chemists by means of a specific example.

Firstly it should be noted that σ and π orbitals are orthogonal for purely symmetry reasons, so that they are treated by separate equations, although terms which allow for interaction between σ and π electrons are included. Any π m.o. will only contain π-type a.o.'s.

For the σ electrons the SCF equation will be for CO

$$\left\{ H^N + \sum_{1}^{5} (2J_{i\sigma} - K_{i\sigma}) + 4J_{1\pi} - 2K_{1\pi} \right\}\sigma = \epsilon^{SCF}\sigma\,. \tag{2}$$

The summation of $i = 1-5$ and the single 1π operator represent the fact that any electron will 'see' charge density due to those electrons which are in the molecule. Now we obtain an m.o. for every a.o. that has been taken in the basis, so that for a double zeta basis set we may have about a dozen m.o.'s. The Hamiltonian is the same, however, for all m.o.'s., both the filled ($i\sigma$ = 1–5) and the unoccupied or virtual orbitals.

For the π orbitals the wave equation in the SCF form is

$$\left\{ H^N + \sum_{i=1}^{5} (2J_{i\sigma} - K_{i\sigma}) + 4J_{1\pi}^0 - K_{1\pi}^2 - K_{1\pi}^0 \right\} \pi = \epsilon_\pi^{SCF} \pi. \quad (3)$$

The two Hamiltonians in (2) and (3) are often written for convenience in a block diagram to facilitate taking matrix elements of the SCF operator. The H^N will always be present and can be left out of the diagram as it is only in the J and K operators that confusion can arise.

The operators of eqns. (2) and (3) can then be written as

Type \ Occupied orbitals	$1-5\sigma^2$	$1\pi^4$
σ	$2J - K$	$4J - 2K$
π	$2J - K$	$4J^0 - K^2 - K^0$

The J and K are 'coulomb' and 'exchange' operators, although in some cases the integrals in $<\phi|H^{SCF}|\phi>$ will not be strictly J or K integrals,

i.e. not $\zeta_{j\sigma j\sigma}^{i\sigma i\sigma}$ and $\zeta_{i\sigma j\sigma}^{i\sigma j\sigma}$, but rather, for example, $\zeta_{m\sigma m\sigma}^{i\sigma j\sigma}$ and $\zeta_{j\sigma m\sigma}^{i\sigma m\sigma}$,

sometimes called pseudo-J or pseudo-K integrals.

The form of the SCF operator is very important, since it enables one to express $<\phi|H^{SCF}|\phi> = \epsilon_\phi^{SCF}$ very simply as a sum of integrals. Thus in our example

$$\epsilon_{1\sigma}^{SCF} = \left\langle 1\sigma \middle| H^N + \sum_{i=1}^{5} (2J - K)_{i\sigma} + 4J_{1\pi} - 2K_{1\pi} \middle| 1\sigma \right\rangle$$

$$= \epsilon_{1\sigma}^N + \sum_{i=1}^{5} (2J - K)_{1\sigma i\sigma} + 4J_{1\pi 1\sigma} - 2K_{1\pi 1\sigma},$$

or, more generally,

$$\epsilon_{i\sigma}^{SCF} = \epsilon_{i\sigma}^N + \sum_{j=1}^{5} (2J - K)_{i\sigma j\sigma} + 4J_{1\pi i\sigma} - 2K_{1\pi i\sigma}$$

and

$$\epsilon_{1\pi}^{SCF} = \epsilon_{1\pi}^N + \sum_{i=1}^{5} (2J - K)_{1\pi i\sigma} + 4J_{1\pi 1\pi}^0 - K_{1\pi 1\pi}^2 - K_{1\pi 1\pi}^0.$$

These orbital-energy expressions may be used to simplify the long energy expression

$$<\Psi|H|\Psi>$$

where Ψ is the Slater-determinant wave function and H the electrostatic Hamiltonian.

33

If we do this, substituting $\epsilon_{i\sigma}^{SCF}$ and $\epsilon_{i\sigma}^{SCF}$, we find

$$E = \sum_{i=1}^{5} (\epsilon_{i\sigma}^{N} + \epsilon_{i\sigma}^{SCF}) + 2(\epsilon_{i\pi}^{N} + \epsilon_{i\pi}^{SCF}),$$

i.e. the energy for a closed-shell molecule is a simple sum of orbital energies and one-electron energies. This result is quite general for all closed-shell molecules, both diatomic and polyatomic.

The technique by which this pleasingly simple result is obtained is very important. Tremendous simplifications can be obtained if one works in terms of ϵ^{SCF} rather than with long lists of integrals derived from use of the electronic Hamiltonian. Some applications of this idea are presented in the next chapter.

6
USES OF ORBITAL ENERGIES

Three applications of the use of SCF orbital energies, ϵ^{SCF}, to reduce energy expressions to simple forms will be given. The technique is of very general use for almost any problem concerning energies. Energy expressions obtained by using the rules for taking matrix elements between Slater determinants are simplified by substituting orbital energy expressions which result from taking matrix elements of the SCF Hamiltonian.

6.1. Ionization potentials

The ϵ^{SCF}, or, as they are often simply called, the orbital energies ϵ, approximate to ionization potentials. To illustrate this let us consider the ionization of CO to CO^+ by removal of a 5σ electron.

The energy expression for CO was given in the last chapter. The wave function for the $^2\Sigma^+$ ground state of the ion will be

$$\Psi = |1\sigma^2 2\sigma^2 3\sigma^2 4\sigma^2 5\sigma\ 1\pi^4|.$$

The energy could be obtained by using the rule for taking matrix elements of the electronic Hamiltonian using Slater's rules and comparing the energy expressions for the molecule and the ion. However, if we assume that the m.o's do not change very much between molecule and ion then many of the terms will be the same in the two expressions. Indeed we see, simply by looking at the two wave functions, that with respect to the neutral molecule the ion will have the same energy expression, except that the following terms will not occur:

$$\epsilon_{5\sigma}^N + \sum_{i=1}^{4} (2J - K)_{5\sigma i\sigma} + J_{5\sigma 5\sigma} + 4J_{5\sigma 1\pi} - 2K_{5\sigma 1\pi},$$

which is identical with $\epsilon_{5\sigma}^{SCF}$.

This result is generally true for closed-shell molecules and is called Koopmans' theorem.

$$I.p.(i) \approx \epsilon_i^{SCF}$$

The table shows how some computed ϵ^{SCF} values agree with orbital energies found from photoelectron spectroscopy. The agreement is such that it is clear that *ab initio* m.o. calculations can be of considerable assistance in interpreting these data.

35

Comparison of Hartree–Fock orbital energies with photoelectron
ionization potentials

Molecule	Orbital	Calc. (eV)	Photoelectron i.p. (vertical)
H_2	$1\sigma_g$	16·18	15·88
HF	1π	17·69	16·06
	3σ	20·90	20·00
CO	4σ	21·87	19·72
	5σ	15·09	14·01
	1π	17·40	16·91
N_2	$3\sigma_g$	17·28	16·96
	$2\sigma_u$	21·17	18·72
O_2	$3\sigma_g$	20·02	20·12
	$1\pi_u$	19·19	17·99
	$1\pi_g$	14·47	12·54

There are great dangers in the indiscriminate application of
Koopmans' theorem. The assumption that all orbitals are unchanged
when going from the molecule to the ion is clearly drastic, but further,
implicit in the derivation of Koopmans' theorem is the notion that the
difference in Hartree–Fock energies of ion and molecule is equal to
the true difference of energies. This is by no means always the case,
due to the inherent errors in the Hartree–Fock method. These errors,
which can even result in predictions of the order of ionization potentials being in error, are discussed in Chapter 10. The classic case of
error is N_2, where in the molecule the $1\pi_u$ electron has a smaller
orbital energy than the $3\sigma_g$ electron, but nonetheless the ground state
of the ion is $^2\Sigma_g^+$ as a consequence of correlation-energy differences in
the ionic states.

6.2. Excitation energies

This same technique of taking matrix elements between determinantal wave functions (remembering as is discussed later that the wave
functions may be linear combinations of determinants) and the use of
ϵ^{SCF} to simplify the expressions can be used to find the excitation energies of other states of a molecule built using virtual orbitals.

Again taking CO as an example; by exciting a 5σ electron to a 6σ
orbital we can obtain a $^3\Sigma^+$ state with the wave function

$$\Psi' = |1\sigma^2 2\sigma^2 3\sigma^2 4\sigma^2 5\sigma\, 6\sigma\, 1\pi^4|.$$

Now the energy of this state could again be found by taking the matrix
element $<\Psi'|H|\Psi'>$ and the excitation energy by subtracting $<\Psi|H|\Psi>$,
the energy of the molecule.

36

This is the safest manner in which to proceed, but it should soon become evident that this long-winded process is by-passed by simply considering gains and losses when going from one state of the molecule to another. In comparing the two expressions, the $^3\Sigma^+$ expression will gain

$$\epsilon_{6\sigma}^{N} + \sum_{i=1}^{4} (2J - K)_{6\sigma i\sigma} + (J - K)_{6\sigma 5\sigma} + (4J - 2K)_{6\sigma 1\pi}$$

and $^1\Sigma^+$ will lose

$$\epsilon_{5\sigma}^{N} + \sum_{i=1}^{4} (2J - K)_{5\sigma i\sigma} + J_{5\sigma 5\sigma} + (4J - 2K)_{5\sigma 1\pi},$$

but

$$\epsilon_{6\sigma}^{SCF} = \epsilon_{6\sigma}^{N} + \sum_{i=1}^{5} (2J - K)_{6\sigma i\sigma} + (4J - 2K)_{6\sigma 1\pi}$$

and

$$\epsilon_{5\sigma}^{SCF} = \epsilon_{5\sigma}^{N} + \sum_{i=1}^{4} (2J - K)_{5\sigma i\sigma} + J_{5\sigma 5\sigma} + (4J - 2K)_{5\sigma 1\pi}$$

$$\therefore\ E(^3\Sigma^+) - E(^1\Sigma^+) = \epsilon_{6\sigma}^{SCF} - \epsilon_{5\sigma}^{SCF} - J_{5\sigma 6\sigma}.$$

In this we have again assumed that most electrons are in identical orbitals in both states of the molecule, but reasonable agreement can nevertheless be achieved, as the following table for CO indicates.

CO excitation energies

Excitation	State	Calculated (eV)		Exp. (eV)
	X $^1\Sigma^+$	0		0
$5\sigma \longrightarrow 2\pi$	$^3\Pi$	6·17	a $^3\Pi$	6·33
$5\sigma \longrightarrow 2\pi$	$^1\Pi$	9·23	A $^1\Pi$	8·51
$1\pi \longrightarrow 2\pi$	$^3\Sigma^+$	7·41	a' $^3\Sigma^+$	8·58
$1\pi \longrightarrow 2\pi$	$^3\Sigma^-$	9·56	I $^1\Sigma^-$	
$1\pi \longrightarrow 2\pi$	$^1\Sigma^-$	9·56	e $^3\Sigma^-$	10·02
$1\pi \longrightarrow 2\pi$	$^3\Delta$	8·48	d $^3\Delta$	9·56

6.3. Configuration interaction

To improve the wave function of a particular state of a molecule it is often necessary to allow a certain mixing with the wave functions of other configurations of the same symmetry. If we consider two states, we could write

$$\Psi = a\Psi_0 + b\Psi_1,$$

where Ψ_0 and Ψ_1 are determinantal wave functions of the same symmetry type and a and b are the mixing coefficients. The best Ψ will be the mixture which gives the lowest energy, so that we have yet another linear variational problem formally just like a Hückel or SCF problem.

We thus have to solve a set of secular equations such as

$$a\,(H_{00} - E) + b\,H_{10} \qquad = 0$$
$$a\,H_{01} \qquad + b(H_{11} - E) = 0,$$

where H_{00} is $\langle\Psi_0|H|\Psi_0\rangle$ and H_{10} $\langle\Psi_{10}|H|\Psi_0\rangle$. This looks exactly like a Hückel calculation, but it must be realised that Ψ_0 and Ψ_1 are complete molecular wave functions, possibly linear combinations of determinants, and the matrix elements involving them must be taken with the aid of the rules given in Chapter 4.

The solution of the secular equations is performed in the usual manner; the secular determinant is set equal to zero and solved to give the possible values of E which may be substituted back into the secular equations to give the mixing coefficients.

Example. Continuing with the CO molecule, we have a ground state $^1\Sigma^+$,

$$\Psi_0 = |1 - 5\sigma^2\,1\pi^4| = A;$$

an excited $^1\Sigma^+$ state,

$$\Psi_1 = \frac{1}{\sqrt{2}}\{|1 - 4\sigma^2 5\sigma\,6\bar{\sigma}\,1\pi^4| - |1 - 4\sigma^2\,5\bar{\sigma}\,6\sigma\,1\pi^4|\}$$

$$= \frac{1}{\sqrt{2}}\,(B - C)$$

The secular determinant will then be

$$\begin{vmatrix} H_{AA} - E & \dfrac{1}{\sqrt{2}}(H_{AB} - H_{AC}) \\[3mm] \dfrac{1}{\sqrt{2}}\,(H_{AB} - H_{AC}) & \dfrac{1}{2}\,(H_{BB} + H_{CC} - 2H_{BC}) - E \end{vmatrix} = 0.$$

The first simplification which can be made to this is to calculate all the diagonal energy terms with respect to the unperturbed energy of the lowest level, i.e. put $H_{AA} = 0$. If the values of all other matrix elements are found using the rules of Chapter 4, it is found that

$$H_{BB} = H_{CC}$$

and $\qquad H_{AB} = -H_{AC}$.

Thus the determinant simplifies to

$$\begin{vmatrix} -E & \sqrt{2}\,H_{AB} \\[3mm] \sqrt{2}\,H_{AB} & (H_{BB} - H_{BC}) - E \end{vmatrix} = 0.$$

From the rule about taking matrix elements between wave functions with one spin orbital different, we see that

$$H_{AB} = \epsilon^N_{5\sigma 6\sigma} + \sum_1^5 \left(2 \zeta^{5\sigma 6\sigma}_{i\sigma i\sigma} - \zeta^{5\sigma i\sigma}_{6\sigma i\sigma} \right) + 4 \zeta^{1\pi 1\pi}_{5\sigma 6\sigma} - 2 \zeta^{5\sigma 1\pi}_{6\sigma 1\pi}.$$

However, using our simplifying trick of using ϵ^{SCF},

$$\epsilon^{SCF}_{5\sigma 6\sigma} = \langle 5\sigma | H^{SCF} | 6\sigma \rangle$$

$$= \epsilon^N_{5\sigma 6\sigma} + \sum_1^5 \left(2 \zeta^{5\sigma 6\sigma}_{i\sigma i\sigma} - \zeta^{5\sigma i\sigma}_{6\sigma i\sigma} \right) + 4 \zeta^{1\pi 1\pi}_{5\sigma 6\sigma} - 2 \zeta^{1\pi 5\sigma}_{1\pi 6\sigma}$$

$$= H_{AB};$$

but the value of non-diagonal matrix elements of the H^{SCF} operator is zero,

$$\text{Therefore } H_{AB} = 0.$$

Thus there is no C.I. between the two states. This is a simple example of the Brillouin theorem. *For closed-shell states of molecules, mono-excited states do not interact directly with the ground state.* However, since mono-excited states can interact with di-excited states, there can be an indirect effect.

Di-excited states may interact with the ground state, as may singly-excited levels of open-shell molecules. For these the technique of configuration interaction followed above is used, but the results may not be so simple.

In general, if a wave function is expressed

$$\Psi = aA + bB + cC + dD + \ldots$$

the C.I. determinant to be solved will be

$$\begin{vmatrix} -E & H_{AB} & H_{AC} & H_{AD} & H_{AD} \ldots \\ H_{BA} & (H_{BB} - H_{AA} - E) & H_{BC} & H_{BD} \ldots \\ H_{CA} & H_{CB} & (H_{CC} - H_{AA} - E) & H_{CD} . \\ . & . & . & . \\ . & . & . & . \end{vmatrix} = 0.$$

The diagonal element components such as $(H_{BB} - H_{AA})$ can be found exactly as excitation energies, to which they are formally identical, and the off-diagonal elements by using Slater's rules for evaluating matrix elements.

6.4. Multiconfigurational SCF procedures

While considering configuration interaction it is convenient to mention 'multiconfiguration SCF', which is a feature of some of the most sophisticated of the *ab initio* computer programmes.[7]

In the C.I. treatment just discussed an improved Ψ was

$$\Psi = a\Psi_0 + b\Psi_1 + c\Psi_2 + \ldots .$$

The variationally computed Ψ_0 is taken, and from it, using the virtual orbitals which result from its computation, several excited states, Ψ_1, etc., of the appropriate symmetry are constructed. The linear variational method is used to find the best possible mixing coefficients a, b, etc.

In multiconfigurational SCF work, a function of the form

$$\Psi = a\Psi_0 + b\Psi_1 + \ldots$$

is found, but not only are the best values of the coefficients a, b, etc., found but also simultaneously the best forms of the constituent wave functions Ψ_1, etc.

In the most sophisticated version complete multiconfigurational SCF wave functions are produced. For a closed-shell molecule all possible di-excitations to the available virtual orbitals may be considered. The resulting wave function is very accurate, but unfortunately demands an enormous amount of computer time.

7
WAVE FUNCTIONS FOR OPEN SHELLS

For closed-shell configurations there is barely any problem at all in writing down the algebraic form of the wave function; it will be a single determinant

$$| \phi_1^2 \; \phi_2^2 \ldots \phi_n^2 \; |.$$

The individual ϕ_i form bases for the representation of the symmetry group of the molecule. In the case of heteronuclear diatomic molecules this is the group $C_{\infty v}$ and for homonuclear molecules $D_{\infty h}$

Given the character table for the molecule it is a simple matter to write down orbitals into which electrons may be fed. This is clearly explained in many texts on quantum mechanics[8] or group theory.[9,10]

For CO or any heteronuclear diatomic molecule these will just be σ orbitals and π orbitals, possibly also δ orbitals for some excited states. If all the orbitals of each symmetry are completely filled the overall symmetry species will be $^1\Sigma^+$, there being no resultant spin or orbital angular momentum, e.g. the ground state of CO

$$| \; 1\sigma^2 \, 2\sigma^2 \, 3\sigma^2 \, 4\sigma^2 \, 5\sigma^2 \, 1\pi^4 | \; .$$

It is when we excite electrons or have open shells that problems may arise e.g. the singlet excited state with a 5σ electron promoted to 6σ must have one of the unpaired electrons of α spin and the other of β spin, the alternative determinants being

$$| 1\sigma^2 \, 2\sigma^2 \, 3\sigma^2 \, 4\sigma^2 \, 5\sigma \quad 6\bar{\sigma} \quad 1\pi^4|$$

or

$$| 1\sigma^2 \, 2\sigma^2 \, 3\sigma^2 \, 4\sigma^2 \, 5\bar{\sigma} \quad 6\sigma \quad 1\pi^4 |$$

Neither of these separately is an eigenfunction of the spin angular momentum operator $\mathcal{S}^2$ which commutes with the Hamiltonian. The actual excited singlet wave function is

$$\frac{1}{\sqrt{2}} \; \{|\text{\textemdash\textemdash}5\sigma \; 6\bar{\sigma} \text{\textemdash} \,| - |\text{\textemdash\textemdash}5\bar{\sigma} \; 6\sigma \text{\textemdash} \,|\} \; .$$

This correct linear combination can be obtained by using step up and step down operators S^+ and S^- to express $\mathscr{S}^2$, but the procedure is difficult for a non-specialist (see Daudel, Lefebvre, and Moser [2]). What we hope to do in this chapter is to give details of a very simple method of constructing spin states from a given configuration, answering questions such as, if we have three electrons in three different orbitals, a, b, c, what are the correct linear combinations of the possible Slater determinants which give doublet functions?

Before treating the problem, however, a few words must be said about the configurations.

7.1. Configurations

To determine the symmetry types, or what spectroscopists call the term manifold, of a diatomic molecule (Σ, Π, or Δ type) for a given electronic configuration, we may first consider that each electron has a set of quantum numbers and then find the resultant, a vector method.

Disregarding spin for the time being, the value of the component of angular momentum along the molecular axis, Λ, can be found by adding the values of m_l for each electron, i.e.

for a σ electron $\quad m_l = 0$
for a π^+ electron $\quad m_l = 1$
for a π^- electron $\quad m_l = -1$

For homonuclear molecules a state will be g or u if there are an even or odd number of u electrons respectively.

To decide whether a Σ state is + or − one tests what happens to the function on reflection in the plane through the nuclei, i.e. changing the azimuthal angle ϕ to $-\phi$, which alters π^+ orbitals to π^- and vice versa.

A few examples should serve to illustrate this procedure, the basis of which is fully treated in a number of text books on quantum mechanics or spectroscopy (see particularly Herzberg [11]).

Configuration	State
σ	Σ
π^+	Π
π^-	Π
$\sigma\sigma$	Σ
$\sigma\pi$	Π
$\pi^+\pi^+$	Δ
$\pi^-\pi^-$	Δ
$\pi^+\pi^-$	Σ
$\pi^+\pi^+\pi^+$	Φ
$\pi^+\pi^+\pi^-$	Π

In all these cases it must be remembered that the Pauli principle applies, so that for instance in the configuration $\pi^+\pi^+\pi^+$ we can only have two orbitals with the same principal quantum number, in which case the electrons must have opposite spins – $1\pi^+ \ 1\bar\pi^+$.

7.2. Spin functions

The problem of spin comes in when we have a configuration that we know gives a state of the desired Λ, but then want the combination of determinants that gives a state of a given spin multiplicity.

To take a particular rather complicated example, if we have the configuration

$$1\pi_u^+ \quad 1\pi_u^- \quad 1\pi_g^+ \quad 1\pi_g^- \quad 3\sigma_g$$

then this is a Σ_g state. Writing the spins underneath,

α	α	α	α	α	– a single determinant giving a $^6\Sigma_g$ states.

α	α	α	α	β	
α	α	α	β	α	These five determinants
α	α	β	α	α	can be combined to give
α	β	α	α	α	four $^4\Sigma_g$ states.
β	α	α	α	α	

α	α	α	β	β	
α	α	β	α	β	
α	β	α	α	β	
β	α	α	α	β	
α	α	β	β	α	These give five $^2\Sigma_g$
α	β	α	β	α	states as linear com-
β	α	α	β	α	binations of the deter-
α	β	β	α	α	minants.
β	α	β	α	α	
β	β	α	α	α	

The question is, what linear combinations of the last ten are doublet functions?

The direct but rather tedious and long-winded method that answers this question is to apply the angular momentum operators directly.[2] In most simple cases this is not necessary, since the resulting functions have been published, e.g. a singlet resulting from two unpaired electrons in orbitals ϕ_a and ϕ_b has the form

$$\frac{1}{\sqrt 2} \ [|-\phi_a \ \bar\phi_b-|-|-\bar\phi_a \ \phi_b-|].$$

Other examples can be found in such works as that of Slater.[8]

A somewhat less tedious approach to the problem is to use projected wave functions, which also have extra advantages. There are several different ways of projecting out the desired spin function starting from a given single determinant. One particular method is that of Nesbet.[12] The authors have found this method very useful in practice and simple in application and suggest that it merits attention and fully justifies the effort required in familiarizing oneself with the manipulation. It is, of course, not the only method available nor is it essential for all applications. It is particularly time-saving, however, in configuration interaction studies.

7.3. Projected spin functions

The details of the mathematical basis of Nesbet's method of producing projected wave functions are complicated by fully explained in his paper.[12] For the present purpose, which is to present a workable scheme to a non-specialist, it will suffice to give an example of the result and to explain its use.

Let us assume we have four orbitals outside the closed-shell part of the molecule, indicated by the dashes

$$\underline{1} \quad \underline{2} \quad \underline{3} \quad \underline{4} \ ;$$

each may be a σ or π^+ orbital, etc. If we want the singlet function, two must be of spin α and two of β. We then have the possible determinants

$$\Phi_A \ = \ |\alpha \ \beta \ \alpha \ \beta|$$
$$\Phi_B \ = \ |\alpha \ \alpha \ \beta \ \beta|$$
$$\Phi_C \ = \ |\beta \ \alpha \ \beta \ \alpha|$$
$$\Phi_D \ = \ |\beta \ \beta \ \alpha \ \alpha|$$
$$\Phi_E \ = \ |\beta \ \alpha \ \alpha \ \beta|$$
$$\Phi_F \ = \ |\alpha \ \beta \ \beta \ \alpha|$$

From these we can derive two projected singlet functions as shown in the following section and Appendix 2.

$$\Psi_I \ = \ B + D - E - F$$
$$\Psi_{II} \ = \ A - \tfrac{1}{2}B + C - \tfrac{1}{2}D - \tfrac{1}{2}E - \tfrac{1}{2}F,$$

where B is shorthand notation for Φ_B, etc.

Nesbet's procedure of gaussian elimination produces these functions relatively easily, but also has an additional bonus in reducing the number of terms involved in matrix elements.

If Ψ_{II} were not a projected wave function

$<\Psi_{II}|H|\Psi_{II}>$ would be $H_{AA} - \frac{1}{2}H_{AB} + H_{AC} - \frac{1}{2}H_{AD} - \frac{1}{2}H_{AE} - \frac{1}{2}H_{AF}$

$$+ \frac{1}{4}H_{BB} - \frac{1}{2}H_{BC} + \frac{1}{4}H_{BD} + \frac{1}{4}H_{BE} + \frac{1}{4}H_{BF}$$

$$+ H_{CC} - \frac{1}{2}H_{CD} - \frac{1}{2}H_{CE} - \frac{1}{2}H_{CF}$$

$$+ \frac{1}{4}H_{DD} + \frac{1}{4}H_{DE} + \frac{1}{4}H_{DF}$$

$$+ \frac{1}{4}H_{EE} + \frac{1}{4}H_{EF}$$

$$+ \frac{1}{4}H_{FF}$$

The result of the Nesbet procedure is to produce the function in a convenient diagram form.

	A	B	C	D	E	F	k_μ
Ψ_I	$\frac{1}{2}$	1*	0	1	-1	-1	4
Ψ_{II}	1*	$-\frac{1}{2}$	1	$-\frac{1}{2}$	$-\frac{1}{2}$	$-\frac{1}{2}$	3

Or generally

	A	B	C	D	E	...	k_μ
Ψ_α	a_{α_1}	a_{α_2}	1*	x_{α_1}	x_{α_2}	...	$k_{\mu\alpha}$
Ψ_β	a_{β_1}	1*	x_{β_1}	x_{β_2}	x_{β_3}	...	$k_{\mu\beta}$

The functions are read from the starred number to the right, including the starred number. the numbers (x) being the coefficients. The $(k_\mu)^{-1/2}$ are normalization constants and numbers (a) to the left of and again including the star are auxiliary coefficients used when taking matrix elements.

If we want the element $<\Psi_\alpha|H|\Psi_\beta>$,

this is equal to $\sqrt{\left(\dfrac{k_{\mu\alpha}}{k_{\mu\beta}}\right)} \left[\sum_i a_\alpha \sum_j x_\beta \; <\Phi_i|H|\Phi_j>\right]$

index i is read from the left of the diagram up to and including the star and j from and including the star to the right.

Example

$$<\Psi_{II}|H|\Psi_{II}> = A\left[A - \frac{1}{2}B + C - \frac{1}{2}D - \frac{1}{2}E - \frac{1}{2}F\right]$$

$$= AA - \frac{1}{2}AB + AC - \frac{1}{2}AD - \frac{1}{2}AE - \frac{1}{2}AF$$

Here $AA \equiv <\Phi_A|H|\Phi_A>$, etc.

Each of these terms AA, AB, etc. must be found using the rules for taking matrix elements between Slater determinants, but clearly there is a great saving of labour.

Similarly,

$$\langle \Psi_I | H | \Psi_I \rangle = \left[\frac{1}{2} A + B \right] [B + D - E - F]$$

$$\langle \Psi_I | H | \Psi_{II} \rangle = \sqrt{\left(\frac{4}{3}\right)} \left[\frac{1}{2} A + B \right] \left[A - \frac{1}{2} B + C - \frac{1}{2} D - \frac{1}{2} E - \frac{1}{2} F \right]$$

or $\langle \Psi_{II} | H | \Psi_I \rangle = \sqrt{\left(\frac{3}{4}\right)} [A] [B + D - E - F]$.

As the last two examples show, there is a great simplification in the process of taking matrix elements once one has the table of coefficients x_j and auxiliary coefficients a_i.

The only problem is the production of the table. Nesbet's most elegant mathematics has provided a simple means of doing this, and for the benefit of the reader tables for all simple cases are included as an appendix to this work.

It should be added that particular care should be taken with Σ states. The Nesbet procedure can only give spin states. When we also need to specify Σ^+ or Σ^- then it may be necessary to add and subtract whole combinations so that the resultant behaves correctly on reflection.

8

OPEN-SHELL SCF METHODS

Calculations on open-shell configurations of molecules, where there are incompletely filled species of molecular orbital, are becoming increasingly important. This is largely due to the fact that most excited states of molecules are open-shell structures and it is in this field where *ab initio* calculations offer the only solution to many of the problems that cannot be resolved by experimental spectroscopy.

There are several techniques for the computation of open-shell wave functions,[13] but only two will be mentioned here. Even for these the reader will be left to look at the original papers for details of the mathematics. The present aim is to present sufficient information for a chemist to use existing programmes, which normally employ one of the two open-shell procedures.

8.1. Roothaan's open-shell method [14]

A rigorous derivation of the Hartree—Fock equations leads to equations of the form

$$H^{\text{SCF}} \phi_i = \sum_j \epsilon_{ij}^{\text{SCF}} \phi_j,$$

the $\epsilon_{ij}^{\text{SCF}}$ being coefficients introduced to ensure the orthogonality of the orbitals ϕ_i. For closed-shell configurations the matrix of ϵ_{ij} can be diagonalized by means of a unitary transformation (effectively this is the same as adding and subtracting rows and columns of a determinant) so that the resulting equations have the simpler form

$$H^{\text{SCF}} \phi_i = \epsilon_i^{\text{SCF}} \phi_i$$

with no 'off-diagonal' multipliers $\epsilon_{ij}^{\text{SCF}}$.

However, for open-shell configurations the general equations cannot necessarily be simplified and the off-diagonal terms remain.

In open-shell configurations there are perhaps doubly (c) and singly occupied (o) orbitals of the same symmetry, which are determined by different sets of SCF equations, whose off-diagonal terms $\epsilon_{ij}^{\text{SCF}}$ cannot be completely eliminated by means of unitary transformations.

However, for a large class of open-shell systems Roothaan has reduced the SCF equations to a single eigenvalue problem by absorbing the off-diagonal terms into the effective Hamiltonian, so that one can obtain an exact solution.

For a general open-shell case, the energy will be

$$E = 2 \underbrace{\sum_k H_k^N + \sum_{kl} (2J - K)_{kl}}_{\text{closed-shell part}} + f \left[\underbrace{2 \sum_m H_m^N + f \sum_{mn} (2aJ - bK)_{mn} +}_{\text{open-shell part}} \right.$$

$$\left. + \underbrace{2 \sum_{km} (2J - K)_{km}}_{\substack{\text{open-closed} \\ \text{interaction}}} \right]$$

Here indices k and l refer to the closed-shell electrons and m and n to the open-shell electrons. a, b, and f are numerical constants which will depend on the particular open-shell problem being considered, f being the fractional occupation of the open shell and a and b differing for different states of the same configuration.

The variation principle can be applied to this expression together with orthogonality constraints and SCF equations of the form

$$F_c \, \phi_c \; = \; \eta_c \, \phi_c$$

$$F_o \, \phi_o \; = \; \eta_o \, \phi_o$$

result. (subscripts c and o refer to closed and open shells)
(It should be noted that the ordering of the η in open-shell calculations is not neccessarily the same as the ordering of the one-electron orbital energies ϵ^{SCF}).

The total energy is then equated to

$$E \; = \; \sum_k (\epsilon_k^N + \eta_k) + f \sum_m (\epsilon_m^N + \eta_m).$$

The operators F_c and F_o contain not only one-electron operators and coulomb and exchange operators as in the closed-shell case, but in addition constants a, b, and f, or expressions involving them. These depend on the open-shell case being considered and must normally be specified when writing the data for an SCF programme.

There is not complete uniformity about the way in which these extra pieces of data a, b, and f or their equivalents are used. To a large extent this depends on the author of the particular programme. However, in the input instructions for all the available programmes there is a table which enables the user to specify the correct combination of constants for his particular case.

The Roothan method can give an exact solution for many open-shell problems and has been widely used. One slight disadvantage is that there will be virtual orbitals for both the closed and open-shell sets so these are not very convenient for configuration interaction, nor is the method applicable to all open-shell configurations. It is also difficult to obtain convergence in SCF iterations when there are two separate Hamiltonians.

8.2.   Nesbet's method [15]

Nesbet's method also involves the use of equivalence restrictions. This means that all orbitals of the same sub-shell, whether paired or unpaired, have the same radial part. By using a single effective Hamiltonian for both open and closed shells we compute a single set of orbitals for both open and closed shells of the same symmetry with off-diagonal ϵ_{ij} being automatically zero. The total wave function is built up from orthonormal doubly and singly occupied orbitals which satisfy the symmetry requirements.

An effective Hamiltonian is written using what may be a quite arbitrary averaging procedure to give a set of SCF equations. The energy will not normally come out to be expressable as a sum of ϵ^N and ϵ^{SCF} terms but should be close to this.

This method is perhaps best understood by considering some simple examples.

As a simple example let us consider HeH, with the configuration $1\sigma^2\ 2\sigma$. We may write the Hamiltonian quite arbitrarily as

$$\{H^N + 2J_{1\sigma} - K_{1\sigma} + J_{2\sigma} - K_{2\sigma}\}.$$

Hence
$$\epsilon_{1\sigma}^{SCF} = \epsilon_{1\sigma}^N + J_{1\sigma1\sigma} + J_{1\sigma2\sigma} - K_{1\sigma2\sigma}$$

and
$$\epsilon_{2\sigma}^{SCF} = \epsilon_{2\sigma}^N + 2J_{1\sigma1\sigma} - K_{1\sigma2\sigma}.$$

The energy
$$E = 2\epsilon_{1\sigma}^N + \epsilon_{2\sigma}^N + J_{1\sigma1\sigma} + 2J_{1\sigma2\sigma} - K_{1\sigma2\sigma},$$

but
$$\left(\epsilon_{1\sigma}^{SCF} + \epsilon_{1\sigma}^N\right) + \frac{1}{2}\left(\epsilon_{2\sigma}^{SCF} + \epsilon_{2\sigma}^N\right) =$$

$$2\epsilon_{1\sigma}^N + J_{1\sigma1\sigma} + J_{1\sigma2\sigma} - K_{1\sigma2\sigma} + \epsilon_{2\sigma}^N + J_{1\sigma2\sigma} - \frac{1}{2}K_{1\sigma2\sigma} = E + \frac{1}{2}K_{1\sigma2\sigma}$$

Now let us consider, as a quite complicated example, an excited state of CO with the wave function

$$| \ 1\sigma^2 2\sigma^2 3\sigma^2 4\sigma^2 5\sigma^2\ 1\pi^+\ 1\overline{\pi}^+\ 1\pi^-\ 6\sigma \ |$$

a single-determinantal $^3\Pi$ level

The general form of the SCF equations given previously is

$$\left\{H^N + \sum_j J_j - \sum_j{}' K_j\right\} \phi_i\ (1) = \epsilon_i^{SCF}\ \phi_i\ (1).$$

For convenience when taking matrix elements the two-electron part of this may again be more specifically written out in a block, but now there is some arbitrariness about this.

Occupied orbitals	$1–5\sigma^2$	6σ	$1\pi^+1\overline{1\pi}^+1\pi^-$
σ	$2J - K$	$J - K$	$3J - 2K$
π^+	$2J - K$	$J - K$	$3J^0 - K^2 - K^0$
$\overline{\pi}^+$	$2J - K$	J	$3J^0 \qquad - K^0$
π^-	$2J - K$	$J - K$	$3J^0 - K^2 - K^0$
Average π electron	$2J - K$	$J - \frac{2}{3}K$	$3J^0 - \frac{2}{3}K^2 - K^0$

Other average Hamiltonians could be written, only each would give a slightly different energy expression. If we set up the SCF equation as just written, how would the energy expression turn out? To answer this we first have to find the energy of the state using the electronic Hamiltonian, i.e. use the rules for taking the matrix element of the wave function with itself.

$$E = <\Psi|H|\Psi>$$

$$= \sum_{i=1}^{5} 2\epsilon_{i\sigma}^N + 3\epsilon_{1\pi}^N + \epsilon_{6\sigma}^N + \sum_{i,j=1}^{5} (2J - K)_{i\sigma j\sigma}$$

$$+ \sum_{i=1}^{5} (6J - 3K)_{i\sigma 1\pi} + \sum_{i=1}^{5} (2J - K)_{6\sigma i\sigma} + 3J_{1\pi 1\pi}^0 - K_{1\pi 1\pi}^2$$

$$+ 3J_{6\sigma 1\pi} - 2K_{6\sigma 1\pi}.$$

We could then use the average SCF Hamiltonian we have just written above to express the ϵ^{SCF} in terms of integrals,

i.e. $\epsilon_{i\sigma}^{SCF} = \epsilon_{i\sigma}^N + \sum_{j=1}^{5} (2J - K)_{i\sigma j\sigma} + J_{6\sigma i\sigma} - K_{6\sigma i\sigma} + (3J - 2K)_{i\sigma 1\pi}$

$\epsilon_{6\sigma}^{SCF} = \epsilon_{6\sigma}^N + \sum_{i=1}^{5} (2J - K)_{i\sigma 6\sigma} + 3J_{1\pi 6\sigma} - 2K_{1\pi 6\sigma}$

$\epsilon_{1\pi}^{SCF} = \epsilon_{1\pi}^N + \sum_{i=1}^{5} (2J - K)_{1\pi i\sigma} + J_{1\pi 6\sigma} - \frac{2}{3}K_{1\pi 6\sigma} + 2J_{1\pi 1\pi}^0 - \frac{2}{3}K_{1\pi 1\pi}$

$\therefore \quad E(^3\Pi) = \sum_{i=1}^{5} (\epsilon^N + \epsilon^{SCF})_{i\sigma} + \frac{1}{2}(\epsilon^N + \epsilon^{SCF})_{6\sigma} + \frac{3}{2}(\epsilon^N + \epsilon^{SCF})_{1\pi} +$

$$+ \frac{1}{2} \sum_{i=1}^{5} K_{6\sigma i\sigma} + \frac{1}{2} \sum_{i=1}^{5} K_{1\pi i\sigma}.$$

Thus the energy towards which we converge is not even the true Hartree–Fock energy since an effective Hamiltonian has been used. However, the corrections to this (in the above case a few exchange integrals K) are small. This means that although the energy is poor when uncorrected, the wave function is close to a true Hartree–Fock solution and sufficiently accurate to compute expectation values including the correction terms to the energy.

This method is extremely simple to use and can be used for any open-shell case. There are also snags here however. The averaging procedure is rather arbitrary and it is not always clear what the best choice is, particularly if the molecular wave function is not a single determinant. Further, the Brillouin theorem no longer holds, so that mono-excited states may not have zero matrix elements with the calculated state even if such matrix elements are zero using the correct Hamiltonian. This means that lengthy configuration-interaction calculations are often necessary.

9

POLYATOMIC MOLECULES

Although diatomic molecules are of interest both in themselves and as
model systems, the chemist will be more concerned with *ab initio*
calculations of polyatomic molecules. Several computer programmes
for carrying out such calculations are in use at the present time, but
before considering the structure and use of these programmes it is
necessary to give a brief account of the configurations of polyatomic
molecules.

9.1. Configurations of polyatomic molecules

An electronic state of a polyatomic molecule is classified according
to the symmetry properties of its wave function. The total electronic
wave function of a given state must transform according to one of the
irreducible representations of the point group of the molecule.

These irreducible representations or symmetry species are normally
labelled using the nomenclature introduced by Placzek and Mulliken.
The symbols A and B represent one-dimensional irreducible representations, the A representation being symmetric and the B representation
antisymmetric with respect to rotation about the principal axis of
symmetry. Two-dimensional representations are labelled E and three-
dimensional representations T or F. A prime ($'$) is added if the species
is symmetric with respect to reflection in a plane of symmetry perpendicular to the principal axis and a double prime ($''$) if it is antisymmetric.
If the molecule possesses a centre of symmetry, the suffixes g and u
are used to distinguish between species which are symmetric and antisymmetric with respect to inversion at this centre, as for homonuclear
diatomic molecules. If there are other elements of symmetry, then the
most symmetric species is given the suffix 1 and the other species
are given the suffixes 2, 3, etc., in decreasing order of symmetry. A
similar system is used in labelling the molecular orbitals, with lower-
case letters instead of upper-case.

In order to determine the symmetry species of the electronic states
that may be obtained from a given molecular orbital configuration it
is necessary to obtain the direct product of the characters of the

molecular orbital species of all the electrons in the molecule. The direct product of the characters is simply the ordered product of the corresponding characters of all the molecular orbital species (here and throughout this chapter we are considering only real representations). The direct product of a given species with itself is always the totally symmetric representation, so closed shells may be disregarded. A completely closed-shell species is automatically the totally symmetric species, necessarily a singlet owing to the Pauli principle.

For open-shell configurations we need only determine the direct product of the partially occupied molecular orbitals. If the configuration contains partially occupied orbitals belonging to two- or three-dimensional symmetry species then the direct product may not be one of the irreducible representations of the point group of the molecule. However, the set of characters of the reducible representation may be decomposed into a linear combination of the characters of the component irreducible representations by inspection of the character table of the group. For example, for a molecule having C_{3v} symmetry and a doubly occupied e orbital states of symmetry A_1, A_2 and E can be obtained. If the e orbital contains three electrons then there is only one state, an E-state, which results. Tables giving the symmetry species of the states which may be obtained from various configurations of molecules belonging to the most typical point groups have been produced by Herzberg. [16]

9.2. Polyatomic calculations

Usually programmes for calculating *ab initio* wave functions for polyatomic molecules consist of three distinct parts which may be used separately or together:
- (a) calculation of all necessary integrals over atomic basis functions,
- (b) construction of integrals over symmetry orbitals (not all programmes include this stage),
- (c) SCF iterations.

Each of these three parts will now be considered in turn.

(a) Calculation of integrals over basis functions

We have already seen for diatomic molecules how the integrals over molecular orbitals required to build the matrix elements of the Hamiltonian prior to iteration are expanded in terms of integrals over atomic basis functions. For polyatomic molecules the same procedure is followed, but in this case the necessary integrals will involve basis functions centred on three or four different atoms. The methods used to calculate the one- and two-centre integrals for diatomic molecules cannot be applied to these multicentre integrals. It has proved difficult to formulate a method to calculate multicentre integrals sufficiently rapidly when the basis functions are the exponential functions (Slater-type orbitals) which are the standard basis functions for calculations on diatomic molecules. One approach to the problem is to use a different type of basis function which is more amenable to integration.

(i) Gaussian functions

The Gaussian function of the form $x^l y^m z^n e^{-ar^2}$, first suggested by Boys,[17] has proved extremely useful in *ab initio* calculations of polyatomic molecules. The product $x^l y^m z^n$ represents the angular distribution of the function. The coefficients l, m, and n can have any integral value. When a basis set of Gaussian functions is used, the necessary multicentre integrals become much simpler to calculate by virtue of the following property: the product of two Gaussian functions G_a and G_b centred on different points a and b is itself a Gaussian function, centred at e somewhere on the line joining these two points. Therefore a three- or four-centre integral may be reduced to a two-centre integral. Thus

$$\left\langle G_a\, G_b \left| \frac{1}{r_{12}} \right| G_c\, G_d \right\rangle \equiv \left\langle G_e \left| \frac{1}{r_{12}} \right| G_f \right\rangle$$

The main disadvantage of the Gaussian function is that it does not resemble very closely the form of real atomic orbital wave functions. In particular, the Gaussian function lacks a cusp at the nucleus and hence the region near the nucleus is described rather poorly unless a large number of functions are used. The behaviour at large distances is also very different from that of the exact atomic orbital wave functions.

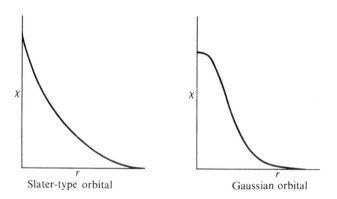

Slater-type orbital   Gaussian orbital

A variational calculation of the energy of the hydrogen atom using a single Gaussian function gives only 4/5 of the total energy, whereas a

single exponential function gives the total energy exactly. This defect
may be overcome by using a large number of Gaussian functions with
suitably chosen exponents in the basis set. However, this introduces
difficulties in the solution of the Roothaan equations in the third part
of the calculation. In particular, it becomes very difficult to get the
iterative process to converge with a very large number of basis
functions. Even if it is possible to obtain convergence, the time required
to build up the matrix elements of the Hamiltonian and diagonalize the
resultant matrix increases enormously if a very large basis set is used.

The problem of the size of the basis set required when Gaussian
functions are used has been studied by Huzinaga.[18] He found that the
number of Gaussian basis functions necessary is more than twice as
great as the number of Slater-type exponential basis functions which
gives an identical energy. This somewhat depressing conclusion
might lead one to believe that accurate calculations using Gaussian
functions would be extremely difficult. However, a method has been
found to reduce the number of variables in the SCF calculation with
very little loss of accuracy. Instead of allowing all the coefficients
of the basis function expansion to vary freely, certain coefficients are
fixed relative to one another, thus forming groups of Gaussian functions,
known as 'contracted Gaussians'.

The m.o. is then expressed as

$$\phi_i = \sum_k c_{ik} \gamma_k,$$

where γ_k is a small contraction of gaussians of the same type on the
same centre,

e.g.
$$\gamma_1 = c_1' \beta_1 + c_2' \beta_2 + c_3' \beta_3$$

In this way, a large basis set may be broken up into a much smaller
number of groups. In the variational calculation of the molecular wave
function only the coefficient of the contracted Gaussian (c) is allowed
to vary and not the relative proportions of the Gaussians within each
group $(c's)$. How much accuracy is lost as a result depends a great
deal on the skill with which the initial basis of Gaussians is contracted.
The contraction process is largely a matter of using chemical intuition.
For example, the 1s orbital of an atom in a molecule is unlikely to
differ greatly from that in the isolated atom, so we can choose a group
of Gaussians having as coefficients those obtained for the 1s orbital
in a variational calculation of the isolated atom. Similarly, in general
it may be expected that only the long-range part of the 2s atomic orbital
will be affected by the deformation caused by the bonding in the
molecule. Therefore one can take another group of Gaussians having as
coefficients those of the 2s orbital in the isolated atom and then

'decomposing' this group into two groups, one containing the Gaussian with the smallest exponent (longest range), and the other group containing all the other Gaussians with the same relative proportions as in the isolated atom. These are just examples of the contraction process. In practice the extent to which the basis set is contracted will depend on the accuracy desired. A set of variational calculations on atoms of the first and second rows and first transition series using Gaussian functions has been published.[18] The coefficients for the contracted Gaussian groups are often taken from this work.

As an example of a contracted Gaussian basis set we take the basis set used by Clementi and Davis[19] for an *ab initio* calculation on the ethane molecule. The exponents and coefficients for this basis set were taken from the set of calculations on the isolated atoms of the first-row elements carried out by Huzinaga.[18] Using a basis set of 10 '1s' and 6 '2p' Gaussian functions, Huzinaga minimized the energy of the carbon atom in the ^3P ground state with respect to variation of the exponents of the Gaussian functions. The exponents and coefficients that he obtained are as follows:

1s

Function	Exponent	Coefficient
β_1	9470·52	0·00045
β_2	1397·56	0·00358
β_3	307·539	0·01934
β_4	85·5419	0·07736
β_5	26·9117	0·22779
β_6	9·4090	0·42695
β_7	3·50002	0·35791
β_8	1·06803	0·04877
β_9	0·400166	−0·00756
β_{10}	0·135124	0·00213

2s

Function	Exponent	Coefficient
β_1	9470·52	−0·00010
β_2	1397·56	−0·00076
β_3	307·539	−0·00418
β_4	85·5419	−0·01701
β_5	26·9117	−0·05399
β_6	9·4090	−0·12134

Contd.

2s Contd.

Function	Exponent	Coefficient
β_7	3·50002	− 0·17554
β_8	1·06803	0·08502
β_9	0·400166	0·60689
β_{10}	0·135124	0·43809

2p

Function	Exponent	Coefficient
β_{11}	25·3655	0·00875
β_{12}	5·77636	0·05479
β_{13}	1·78730	0·18263
β_{14}	0·65771	0·35871
β_{15}	0·24805	0·43276
β_{16}	0·091064	0·20347

Clementi and Davis grouped these functions together in the following way. To represent the region near the nucleus, which is poorly described by a single Gaussian function, they grouped together the five functions $\beta_1 - \beta_5$, all of which have rather large exponents. To represent the remainder of the 1s orbital they grouped together the two functions β_6 and β_7, which have the largest coefficients in the expansion of the 1s orbital in the atom. The 2s orbital, other than the region close to the nucleus, was represented by the three functions β_8, β_9, and β_{10}, which have large coefficients in the 2s orbital expansion. They did not contract these three functions into a single group but separated the function with the smallest exponent, β_{10}, from the pair of functions β_8 and β_9. In this way they partially allowed for the distortion of the 2s orbital upon bond formation. The six 2p functions were divided into two groups, a group formed from the four functions $\beta_{11} - \beta_{14}$ and a group consisting of the pair of functions with the smallest coefficients, again to make some allowance for the distortion of the 2p orbital upon formation of the molecule. Within each group of contracted Gaussian functions the coefficients used were those obtained in the atomic calculation. The coefficients for the pair of functions β_8 and β_9 were taken from the 1s orbital expansion. The 2p$_x$, 2p$_y$, and 2p$_z$ orbitals had identical exponents and coefficients.

We therefore have the following groups of contracted Gaussian orbitals for the carbon atom:

Group 1 : $0·00045 \, \beta_1 + 0·00358 \, \beta_2 + 0·01934 \, \beta_3 +$
$+ \; 0·07736 \, \beta_4 + 0·22779 \, \beta_5$

Group 2 : $0.42695\ \beta_6 + 0.35791\ \beta_1$

Group 3 : $0.08502\ \beta_8 + 0.60689\ \beta_9$

Group 4 : β_{10}

Group 5 : $0.35871\ \beta_{11} + 0.18263\ \beta_{12} + 0.05479\ \beta_{13} + 0.00875\ \beta_{14}$

Group 6 : $0.43276\ \beta_{15} + 0.20347\ \beta_{16}$

The exponents and coefficients for the hydrogen atom were also taken from the atomic calculations of Huzinaga. They were taken from a four-term expansion in Gaussian orbitals of a hydrogenic 1s orbital with exponent 1.0.

Exponent	Coefficient
0.123317	0.50907
0.453757	0.47449
2.01330	0.13424
13.3615	0.01906

To this expansion Clementi and Davis added a fifth Gaussian function with exponent 0.07983. The four-term expansion formed one contracted Gaussian group and the additional function was allowed to vary freely. More recently it has been found that the exponents of a group of Gaussian functions representing a hydrogen orbital should be multiplied by a scale factor to take into account the contraction of this orbital in the molecular environment. A suitable scale factor for a single Slater-type orbital has been found to be 1.4 for a hydrogen atom bonded to a first-row atom. If the hydrogen orbital is represented by a group of Gaussian functions this means that the exponent of every member of the group should be multiplied by a scale factor of 2.

Another set of contracted Gaussians that has been used with considerable success is the 'Gaussian-lobe' set proposed by Whitten.[20] The Gaussian-lobe functions avoid the computational difficulties involved in integrating over the angular part of the basis functions by expanding the functions that have angular dependence as linear combinations of simple Gaussians without an angular part which are centred at different points in space. Thus a p orbital is represented by a linear combination of two 'lobes' consisting of Gaussians with radial dependence only, the centres of which are displaced an equal distance above and below the atomic nucleus, respectively.

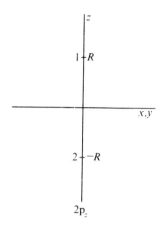

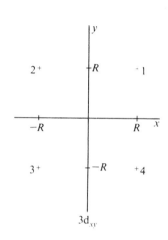

(ii) Slater-type orbitals.

Because of the disadvantage of Gaussians, work has continued on methods of calculating polyatomic molecules using Slater-type orbitals, despite the computational difficulties. The problem of multicentre integrals may be avoided completely by restricting the basis functions to those on a single centre. The m.o.s on other centres are expanded in terms of basis functions on the first centre. Some success has been obtained with this type of basis set in calculations on the hydrides of the elements of the first row, but the method does not seem to be readily extendable to larger molecules. The difficulty is that it is virtually impossible to expand the inner shell around one atom in terms of Slater-type orbitals centred on another atom.

The first, and for a long time the only general method for evaluating multicentre integrals over exponential functions was that due to Barnett and Coulson.[21] In this method the exponential functions on the different centres are all expanded as a power series, known as the 'zeta-function expansion' about a single centre. The necessary integrals are then obtained by integrating the expansion term-by-term. This reduces the calculation of a multicentre integral to the calculation of a large number of single-centre integrals.

The Gaussian transform method of Shavitt and Karplus[22] enables calculations to be carried out using a basis set of Slater-type orbitals while taking advantage of the special properties of the Gaussian function. The method makes use of the integral transform:

$$e^{-ar} = \frac{a}{2\sqrt{\pi}} \int_0^\infty s^{-3/2} \, e^{-a^2/4s} \, e^{-sr^2} \, ds.$$

After transformation of the exponential basis functions, the usual Gaussian formulae may be used to evaluate the integrals, but there is an additional integration over the variable s. However, with efficient programming the extra variable need not be a great drawback. This appears to be one of the most promising methods for the evaluation of multicentre integrals at the present time.

For the first part of the calculation the programme will require as input data information about the geometry of the molecule in the form of coordinates of each atomic centre and information about the the basis functions, in particular the orbital exponents and the atomic centre to which each basis function belongs. If contracted Gaussians are being used, the coefficients of the Gaussians in each group must also be supplied. If the molecule possesses some symmetry, then further information should be supplied to prevent unnecessary calculation of integrals that are equal to other integrals by symmetry or that are zero by symmetry. For a symmetric molecule the calculation time can be halved by including in the programme a procedure to avoid such unnecessary calculation of integrals. However, such a procedure has not proved easy to formulate. Some programmes require a decision prior to the calculation as to which blocks of integrals will not be required, this information then being given in the input data. Another method, useful when contracted Gaussians are being used, involves the calculation of a set of model integrals for a molecule of the required symmetry. This calculation can be carried out very rapidly by using only one Gaussian per group of Gaussians in the full calculation. These model integrals are written onto magnetic tape and classified as zero or equal by symmetry using a separate programme, the classification information again being written on magnetic tape. This information is then read by the programme in the full calculation and used to decide whether a given integral is to be calculated or not, as there will be a one-to-one correspondence between the order of the model integrals and the order of calculation of the 'true' integrals. This method, though somewhat lengthy, is less liable to human error than the first method. The most recent programmes for *ab initio* calculations of polyatomic molecules require only some information about the symmetry operations belonging to the point group of the molecule and the results of these symmetry

60

operations on the different atomic centres. On the basis of this information the programme itself calculates which integrals will not be
required.

The necessary integrals having been calculated, they are written in
a predetermined order in blocks of about 1000 on some suitable storage
device, usually a magnetic tape (if contracted Gaussians are being used
only the integrals over groups are written out, not the integrals over the
individual components of the group).

(b) *Calculation of integrals over symmetry orbitals*

In some programmes this is not a separate step but is carried out
simultaneously with the building up of the matrix elements of the
Hamiltonian. This step consists simply of forming the appropriate
linear combinations of integrals over basis functions to obtain the same
number of integrals over symmetry orbitals.

Although in theory this is a simple operation, in practice it is
difficult to carry out efficiently, as a large amount of time can be
wasted simply manipulating the magnetic tape that contains the integrals
over basis functions in order to pick up a given integral. It is often
very desirable to carry out this step, however, as the use of symmetry
properties can simplify enormously the third step, the SCF iterations,
as explained in Chapter 1. The gain in time in stage 3 has to be balanced
against the time taken in the transformation. The programme will require
as input for this stage the coefficients of the linear combination of
basis functions needed to form each symmetry orbital. These coefficients
may often be obtained simply by inspection as for H_2O in the example
to follow, or from one of the works on the applications of group theory
in chemistry. Alternatively, for difficult cases the symmetry orbital
coefficients may be obtained systematically by means of certain
operators known as projection operators, which project out of any
function that part of it belonging to a given irreducible representation.
The application of these operators is again shown by means of an
example.

Particular care must be taken when dealing with degenerate orbitals
such as π or e orbitals. The symmetry orbitals supplied to the programme
must belong to a given row of the degenerate representation. Functions
belonging to different rows of a representation are normally treated
quite separately by the programme. SCF matrices are set up and
diagonalized for each component of a degenerate e orbital just as if they
belonged to quite different irreducible representations such as a or b.

(c) *SCF calculation*

In the third and final part of the calculation the Hartree–Fock
matrix for each representation is built up and the resultant matrices are

separately diagonalized. The eigenvectors obtained are used as input for the next iteration and the process is repeated until the input and output eigenvectors agree within a certain threshold (usually $\sim 10^{-6}$). As explained in Chapter 3, the initial Hartree—Fock matrices are built up using a guessed set of coefficients, provided as input to the programme. In addition the programme also needs the number of different m.o. symmetry types for the molecule and how many m.o.s of each symmetry type are occupied.

In order to obtain convergence in the SCF iterations the initial coefficients must be reasonable approximations to the final vectors, except for small, highly symmetric molecules. The process of convergence can be greatly speeded up by means of an extrapolation procedure which uses three sets of eigenvectors from two successive iterations to obtain a closer approximation to the final result. In some cases such a procedure is essential in order to obtain convergence at all.

The final wave function is printed out in a convenient form and is also punched out on cards in order to be used as input for other programmes such as those to carry out C.I. calculations or to calculate the dipole moment and other one-electron properties. Some programmes contain procedures to carry out these calculations automatically at the end of each SCF calculation if so desired.

10

EXAMPLES OF POLYATOMIC CALCULATIONS

Although input requirements vary to some extent from one programme to another, it is clear that in most cases a certain amount of information about the molecule must be obtained before a calculation can begin. It is usually not sufficient simply to give the geometry of the molecule and let the programme get on with the calculation. We now consider in more detail the information that is required and the methods by which it is obtained by means of some simple examples. The molecules chosen as examples are probably much simpler than the molecules that interest the majority of chemists but we wish to illustrate the sort of approach that should be employed rather than discuss in detail the necessary group theory techniques, for which a number of excellent texts already exist.[9,10] The way in which a calculation of a more complex molecule is approached is essentially the same as that described here and no additional techniques are required.

10.1. H_2O

We shall take as our first example of a calculation of a polyatomic molecule the very simple case of H_2O. In its equilibrium configuration the water molecule possesses a two-fold axis of symmetry along the bisector of the apex angle and two perpendicular planes of symmetry passing through this axis. Hence the molecule belongs to the point group C_{2v}. In order to carry out the calculation as efficiently as possible we require the symmetry orbitals for H_2O in the group C_{2v} and the electronic configuration of the molecule with the molecular orbitals classified according to their irreducible representation in C_{2v}. With a knowledge of the character table of the group C_{2v} this information may readily be obtained by observing the way in which the atomic orbitals on each of the atomic centres transform under the various operations of the group.

The character table for the group C_{2v} is as follows:

	E	C_2	$\sigma_v(xz)$	$\sigma_v(yz)$
A_1	1	1	1	1
A_2	1	1	-1	-1
B_1	1	-1	1	-1
B_2	1	-1	-1	1

The system of axes is chosen according to the usual convention, with the C_2 axis of symmetry as the z axis and the plane of the molecule as the yz plane.

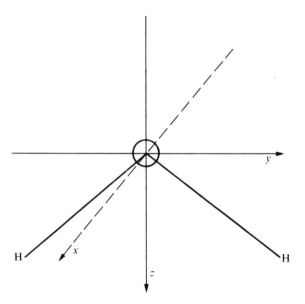

Let us consider first the atomic orbitals centred on the oxygen atom, namely $1s^0$, $2s^0$, $2p_x^0$, $2p_y^0$, and $2p_z^0$. It may be seen at once that the $1s^0$ and $2s^0$ orbitals are not affected by any of the operations of the group C_{2v}. Each of these will therefore be a symmetry orbital belonging to the irreducible representation A_1. Further inspection shows that the $2p_z^0$ orbital is similarly left unchanged by all the operations of the group. The $2p_x^0$ orbital is inverted by the operation of rotation by $180°$ about the two-fold axis of symmetry (C_2) and by reflection in the plane of the molecule $[\sigma(yz)]$. It therefore belongs to the representation B_1. The $2p_y^0$ orbital is also inverted by the C_2 operation and by reflection in the plane through the symmetry axis perpendicular to the plane of the molecule $[\sigma(xz)]$. This orbital transforms as the irreducible representation B_2. Now we consider the orbitals centred on the hydrogen atoms $1s_1^H$

64

and $1s^H$. It is evident that neither of these orbitals by itself is a symmetry orbital of the group C_{2v} since under the operations C_2 and $\sigma(xz)$, $1s_1^H$ transforms into $1s_2^H$ and vice versa. The correct linear combinations are easily found, however. The (unnormalized) combination $(1s_1^H + 1s_2^H)$ will transform as the irreducible representation A_1 and the antisymmetric combination $(1s_1^H - 1s_2^H)$ will transform as the irreducible representation B_2.

Summarizing, the symmetry orbitals that we have obtained for H_2O are

$$A_1 \begin{cases} 1s^0 \\ 2s^0 \\ 2p_z^0 \\ (1s_1^H + 1s_2^H) \end{cases}$$

$$B_2 \begin{cases} 2p_y^0 \\ (1s_1^H - 1s_2^H) \end{cases}$$

$$B_1 \quad 2p_x^0 .$$

The programme will therefore build up matrices for each of these representations and diagonalize them separately. From each matrix a set of orbital energies will be obtained. However, in order to carry this out the programme will require to know how many orbitals belonging to a given irreducible representation are occupied (and the occupancy of each orbital if there are open shells).

The lowest-lying molecular orbital in H_2O will be obviously largely a 1s orbital on the oxygen atom and will therefore have a_1 symmetry. The two symmetry orbitals of B_2 symmetry will give rise to a bonding and antibonding pair of molecular orbitals. Similarly, neglecting 2s–2p hybridization, the 2s and $2p_z^0$ orbitals will separately form bonding and antibonding pairs of molecular orbitals of a_1 symmetry with the symmetric combination $(1s_1^H + 1s_2^H)$. In fact there will be some 2s–2p hybridization, but this will not alter the general picture. The $2p_x^0$ orbital will give a non-bonding orbital of symmetry b_1 as it is the only atomic orbital of this symmetry. H_2O is a ten-electron molecule so, filling the bonding and non-bonding molecular orbitals, there will be three occupied orbitals of a_1 symmetry and one occupied orbital belonging to each of the representations b_2 and b_1. It is useful to have some idea of the composition and relative energy of these orbitals in order to ensure that the SCF iterations converge, although this information is probably not essential for such a simple molecule as H_2O. As stated above, the $1a_1$ orbital will be largely a 1s oxygen atomic orbital and will lie far below the others in energy (~ 20 a.u.). Neglecting 2s–2p hybridization, the $2a_1$ orbital will be largely $2s^0$ and the $3a_1$ largely $2p_z^0$. The $1b_2$ orbital will be mainly $2p_y^0$ and the $1b_1$ orbital will

will be the non-bonding $2p_x^0$. The $1b_2$ orbital may be predicted to lie slightly below the $3a_1$ orbital in energy as the overlap of the $2p_y^0$ orbital with the hydrogen 1s orbitals would be expected to be somewhat greater than that of $2p_z^0$ at the equilibrium configuration (bond angle $105°$). The electronic configuration of H_2O, writing the orbitals in order of increasing energy is therefore

$$(1a_1)^2 \ (2a_1)^2 \ (1b_2)^2 \ (3a_1)^2 \ (1b_1)^2 .$$

The total energy of this configuration expressed as a sum of one-electron terms and coulomb and exchange integrals over molecular orbitals may be obtained using the rules for calculating the matrix elements between Slater determinants given in Chapter 4 and several times illustrated for diatomic molecules.

$$E = \left\langle |1a_1^2 \, 2a_1^2 \, 1b_2^2 \, 3a_1^2 \, 1b_1^2 |H| \, 1a_1^2 \, 2a_1^2 \, 1b_2^2 \, 3a_1^2 \, 1b_1^2| \right\rangle$$

$$= \sum_{i-1}^{3} 2\epsilon_{ia_1}^{N} + 2\epsilon_{1b_2}^{N} + 2\epsilon_{1b_1}^{N} + \sum_{i,j=1}^{3} (2J - K)_{ia_1 ja_1} +$$

$$+ 2\sum_{i=1}^{3}(2J - K)_{ia_1 1b_2} + 2\sum_{i=1}^{3}(2J - K)_{ia_1 1b_1} + 2(2J - K)_{1b_1 1b_2}$$

Using the nomenclature introduced earlier for diatomic molecules, the SCF equations will have the form (λ signifies any one of the symmetry species of the molecule)

$$\left\{ H_\lambda^N + \sum_{i=1}^{3} (2J_{ia_1} - K_{ia_1}) + (2J_{1b_1} - K_{1b_1}) + (2J_{1b_2} - K_{1b_2}) \right\} \phi_\lambda = \epsilon^{SCF} \phi_\lambda .$$

Expressing the orbital energies as a sum of integrals, as before,

$$\epsilon_{ia_1}^{SCF} = \left\langle ia_1 | H_{ia_1}^{N} + \sum_{j=1}^{3} (2J_{ja_1} - K_{ja_1}) + (2J_{1b_2} - K_{1b_2}) + \right.$$
$$\left. + (2J_{1b_1} - K_{1b_1}) | ia_1 \right\rangle$$

$$= \epsilon_{ia_1}^{N} + \sum_{j=1}^{3} (2J - K)_{ia_1 ja_1} + (2J - K)_{ia_1 1b_2} + (2J - K)_{ia_1 1b_1}$$

and

$$\epsilon_{1b_2}^{SCF} = \left\langle 1b_2 | H_{1b_2}^{N} + \sum_{i=1}^{3} (2J_{ia_1} - K_{ia_1}) + (2J_{1b_2} - K_{1b_2}) + \right.$$
$$\left. + (2J_{1b_1} - K_{1b_1}) | 1b_2 \right\rangle$$

$$= \epsilon_{1b_2}^{N} + \sum_{i=1}^{3} (2J - K)_{ia_1 1b_2} + (2J - K)_{1b_2 1b_2} + (2J - K)_{1b_1 1b_2}$$

and

$$\epsilon_{1b_1}^{SCF} = \left\langle 1b_1 | H_{ib_1}^{N} + \sum_{i=1}^{3} (2J_{ia_1} - K_{ia_1}) + (2J_{1b_2} - K_{1b_2}) + \right.$$
$$\left. + (2J_{1b_1} - K_{1b_1}) | 1b_1 \right\rangle$$

$$= \epsilon_{1b_1}^{N} + \sum_{i=1}^{3} (2J - K)_{ia_11b_1} + (2J - K)_{1b_11b_2} + (2J - K)_{1b_11b_1}$$

Therefore, expressing the total energy as a sum of orbital energies, and one-electron energies,

$$E = \sum_{i=1}^{3} (\epsilon_{ia_1}^{N} + \epsilon_{ia_1}^{SCF}) + (\epsilon_{1b_2}^{N} + \epsilon_{1b_2}^{SCF}) + (\epsilon_{1b_1}^{N} + \epsilon_{1b_1}^{SCF})$$

This again illustrates the general result for closed-shell molecules.

10.2. NH_3

The second example we take is the ammonia molecule in the pyramidal configuration. This example will serve to introduce certain more systematic techniques for obtaining the irreducible representations of the symmetry orbitals and the correct linear combinations of basis orbitals that form the symmetry orbitals. The point group of NH_3 in the pyramidal configuration is C_{3v}. The character table of this group is given below.

	E	$2C_3$	$3\sigma_v$
A_1	1	1	1
A_2	1	1	-1
E	2	-1	0

The z axis is taken to be the C_3 axis of symmetry. The $z-x$ plane includes one of the N–H bonds.

The diagram below represents a view of the three hydrogen atoms, looking down the z axis.

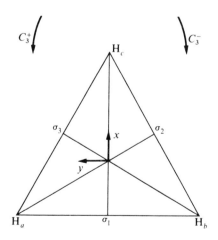

In this case, the symmetry orbital coefficients cannot be obtained simply by inspection as for H_2O because of the two-dimensional E representation, nor is it obvious which atomic orbitals will form the basis for a given irreducible representation.

To determine the irreducible representation of the symmetry orbitals that can be obtained from a set of symmetry-related atomic orbitals the following procedure is carried out. Apply one of the operations of the group to each of the atomic orbitals of the set in turn. The result of the operation on each orbital will be a linear combination of all the orbitals in the set (some of which will have zero coefficients). Express the set of coefficients obtained in this way in matrix form, thus obtaining the representative of this operation and determine the trace of the matrix (the character of the representative). Repeat this process for all the operations of the group. The resultant set of characters may usually be 'decomposed' by inspection into the characters of its component irreducible representations.

There is usually no need to carry out the whole of this procedure. Very often, on applying an operation, the basis function will be completely unaffected or will simply change sign, thus contributing 1 or -1 to the trace, respectively. The operation may also transform the function entirely into other functions of the set and so for this function the contribution to the trace will be zero. For example, the three symmetry-related 1s hydrogen orbitals are all left unchanged by the identity operation E, and so the character for this operation will be 3. On applying the rotation operation C_3, each of the hydrogen orbitals is transformed into another one of the set, so the character will be zero. Each of the reflection operations σ_v will leave one of the orbitals unchanged and interchange the other two, giving a character of 1. We obtain, therefore, the following set of characters.

E	$2C_3$	$3\sigma_v$
3	0	1

We must now look for the linear combination of the characters of the irreducible representations of the group C_{3v} that will give this particular set of characters. It is readily seen from the character table that the desired combination may be obtained by addition of the characters of the representation E and the representation A_1. These will therefore be the irreducible representations of the symmetry orbitals which may be obtained from the three 1s orbitals. In more complicated cases it may not be possible to obtain the component irreducible representations simply by inspection of the character table of the group. They may be obtained systematically using the formula

$$a_j = \frac{1}{h} \sum_R \chi(R)\ \chi_j(R),$$

where a_j is the number of times the jth irreducible representation is contained in the reducible representation (which must be real).

$\chi(R)$ is the character of the reducible representation for the operation R and $\chi_j(R)$ is the character of the jth irreducible representation for the same operation.

h is the order (number of operations) of the group.

The derivation of this formula may be found in any of the references given at the beginning of this section. Instead of decomposing the set of characters obtained for the three 1s hydrogen orbitals by inspection, the irreducible representations could have been obtained by substitution in the above formula. Considering first the A_1 irreducible representation, we obtain

$$a_{A_1} = \tfrac{1}{6}\{3 \times 1 + 2(1 \times 0) + 3(1 \times 1)\} = 1.$$

Similarly, for the E irreducible representation

$$a_E = \tfrac{1}{6}\{3 \times 2 + 2(0 \times -1) + 3(1 \times 0)\} = 1.$$

Thus the A_1 representation and E representation each occur once in the reducible representation having the set of characters obtained above. There is no need to continue further as these two representations combined give a reducible representation with the correct number of dimensions (3).

The orbitals centred on the nitrogen atom are more easily grouped into their respective irreducible representations. The 1s and 2s atomic orbitals are left unchanged by any operation of the group so they belong to the totally symmetric A_1 representation. Because of our particular choice of z axis the $2p_z$ orbital is also left invariant under any operation of the group. The $2p_x$ and $2p_y$ orbitals, on the other hand, are transformed by the C_3 rotation into a linear combination of $2p_x$ and $2p_y$. They must therefore form a basis for the two-dimensional E representation.

We must now obtain the coefficients of the linear combinations of atomic orbitals which form the symmetry orbitals of the group. When these coefficients are not obvious by inspection they may be obtained systematically by means of certain operators known as projection operators which project out of any function that part of it belonging to a given irreducible representation. These operators have the form

$$P(j) = \frac{l_i}{h} \sum_R \chi_j(R)\,R,$$

where R is one of the operations of the point group of the molecule.

$\chi_j(R)$ is the character of the operation R in the jth irreducible representation. (We again assume that we are considering only real representations).

l_j is the dimension of the jth representation and h is the order (number of operations) of the group. The procedure to obtain the symmetry orbital coefficients of the jth irreducible representation will therefore be as follows: choose one of the basis functions and perform each of the operations of the point group of the molecule on this function in turn. Multiply each of the resulting functions in turn by the character of the jth irreducible representation corresponding to the operation which was used to obtain that function. The sum of the functions obtained in this way is a symmetry orbital of the jth irreducible representation. The numerical factor introduced by the term l_j/h may be omitted as it is replaced by a normalization coefficient by the programme. The above procedure will give only one of the components of the set of symmetry orbitals belonging to a degenerate representation. To obtain the other components the process must be repeated for a number of basis functions at least equal to the dimension of the representation. There is a further difficulty for degenerate representations because this method does not in general yield functions that belong to a given row of a degenerate representation, which are the functions that must be supplied to the programme. This difficulty can be overcome by deducing the full matrix of the degenerate representation for any one of the operations of the group. (See in particular Altmann [23].)

Let us consider once more the 1s orbitals on the three hydrogen atoms of the ammonia molecule. We have seen that from these functions symmetry orbitals belonging to the A_1 and E irreducible representations may be formed. We must now determine the linear combinations of the hydrogen orbitals which belong to these two irreducible representations. Each 1s orbital is represented by a single basis function labelled $1s_A$, $1s_B$, and $1s_C$ respectively. We use the character table for the group C_{3v}, given above. Applying the operator for the representation A_1 to basis function $1s_A$, we obtain the symmetry orbitals

$$\Psi(A_1) = 1s_A + 1s_B + 1s_C.$$

Applying the projection operator for the E_1 representation to each of the basis functions in turn yields the functions

$$\Psi_1(E) = 2(1s_A) - 1s_B - 1s_C$$
$$\Psi_2(E) = 2(1s_B) - 1s_A - 1s_C$$
$$\Psi_3(E) = 2(1s_C) - 1s_B - 1s_A.$$

These three functions are not linearly independent. We must find two linear combinations of the three functions which will correspond to different rows of the E representation. Let us consider the matrix corresponding to one of three reflection operations (σ_1, σ_2, σ_3). This matrix may be put in diagonal form. As the character of this operation

70

is zero in the E representation the diagonal matrix must be $\begin{pmatrix} 1 & \\ & -1 \end{pmatrix}$.
Therefore, one of the required functions will be symmetric with respect
to any one of the reflection operations, σ_1, for example, and the other
antisymmetric. The symmetric function is just the third of the three
functions obtained with the projection operator, namely

$$\Psi_1'(E) = 2(1s_C) - 1s_B - 1s_A .$$

The antisymmetric function is the difference of the first and second
functions

$$\Psi_2'(E) = \left(2(1s_A) - 1s_B - 1s_C\right) - \left(2(1s_B) - 1s_A - 1s_C\right)$$

$$= 3(1s_A - 1s_B).$$

The numerical factor may be discarded. Normalization of these functions
is usually carried out by the programme.

The $2p_x$ and $2p_y$ orbitals on the nitrogen atom also form a basis for
the E representation. The σ_1 plane of reflection contains the x axis. Hence
the $2p_x$ orbital will be symmetric with respect to the reflection σ_1 and
the $2p_y$ orbital will be antisymmetric. The $2p_x$ orbital therefore belongs
with the first of the pair of functions obtained above and the $2p_y$ orbital
with the second.

The symmetry orbitals for NH_3 may be summarized as follows:

$$A_1 \quad \begin{matrix} 1s^N \\ 2s^N \\ 2p_z^N \\ 1s_A^H + 1s_B^H + 1s_C^H \end{matrix}$$

$$E \quad \begin{cases} 2(1s_C^H) - 1s_B^H - 1s_A^H \\ 2p_x^N \end{cases}$$

$$\begin{cases} 1s_A^H - 1s_B^H \\ 2p_y^N \end{cases}$$

The symmetry orbitals of E symmetry will give one bonding e orbital.
The other occupied molecular orbitals will have a_1 symmetry. Reasonable guesses of the starting coefficients can be made by assuming that
the $1a_1$ orbital is almost entirely a nitrogen 1s orbital and the $2a_1$ and
$3a_1$ orbitals are predominantly nitrogen 2s and $2p_z$ respectively, together
with a small amount of the hydrogen A_1 symmetry orbital. Assuming that
the overlap of the $2p_x$ and $2p_y$ orbitals with the hydrogen orbitals is
slightly greater than that of the $2p_z$ orbital, the electronic configuration
of NH_3 is:

$$(1a_1)^2 \ (2a_1)^2 \ (1e)^4 \ (3a_1)^2$$

The expression for the total energy of this configuration may be
obtained in the same way as for H_2O, using Slater's rules.

10.3. MF_6

As a final example we shall consider the hexafluoride of a transition metal M in an octahedral configuration. An *ab initio* calculation of such a large molecule might appear a somewhat ambitious project. However, a number of such calculations have already been carried out on this type of molecule.[34]

The coordinate system that we shall use is shown in the diagram below.

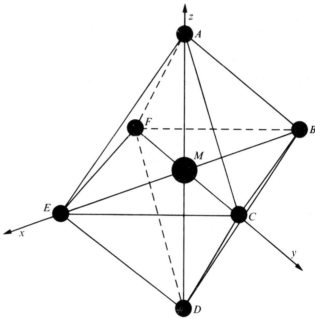

The point group of a molecule in an octahedral configuration is O_h. To determine the symmetry orbitals for this point group we shall need the character table of the group, which is given below.

O_h	E	$8C_3$	$3C_2$	$6C_4$	$6C_2'$	i	$8S_6$	$3\sigma_h$	$6S_4$	$6\sigma_d$
A_{1g}	1	1	1	1	1	1	1	1	1	1
A_{1u}	1	1	1	1	1	−1	−1	−1	−1	−1
A_{2g}	1	1	1	−1	−1	1	1	1	−1	−1
A_{2u}	1	1	1	−1	−1	−1	−1	−1	1	1
E_g	2	−1	2	0	0	2	−1	2	0	0
E_u	2	−1	2	0	0	−2	1	−2	0	0
T_{1g}	3	0	−1	1	−1	3	0	−1	1	−1
T_{1u}	3	0	−1	1	−1	−3	0	1	−1	1
T_{2g}	3	0	−1	−1	1	3	0	−1	−1	1
T_{2u}	3	0	−1	−1	1	−3	0	1	1	−1

The determination of the symmetry orbitals in this case will be a much longer process than it was for the simple molecules we have just considered, but the same methods can be used.

We consider first the atomic orbitals centred on the transition metal M. The orbitals of s symmetry are obviously invariant under all the operations of the group and hence belong to the totally symmetric A_{1g} irreducible representation. However, it is not obvious to which irreducible representations the p and d orbitals belong. Let us take the p orbitals first. To determine the irreducible representations for which the p orbitals form a basis we operate on all the p orbitals with each of the operations of the group in turn and determine the trace of the matrix corresponding to each operation. Instead of treating each type of rotation operation as a separate case we shall consider first the general rotation by an arbitrary angle α.

The axis of rotation is chosen to be the z axis, so the p_z orbital is left unchanged by any rotation about this axis. The p_x and p_y orbitals are transformed into linear combinations of p_x and p_y given by the following expressions.

$$C(\alpha)\,p_x = \cos\alpha\,p_x - \sin\alpha\,p_y$$
$$C(\alpha)\,p_y = \cos\alpha\,p_x + \sin\alpha\,p_y.$$

The result of the operation $C(\alpha)$ on the p orbitals may therefore be expressed in matrix form as follows.

$$C(\alpha)\,\widetilde{p_x p_y p_z} = \widetilde{p_x p_y p_z} \begin{pmatrix} \cos\alpha & \sin\alpha & 0 \\ -\sin\alpha & \cos\alpha & 0 \\ 0 & 0 & 1 \end{pmatrix}$$

The character, $\chi[C(\alpha)]$, of the representation for which the p orbitals from a basis is therefore $(1 + 2\cos\alpha)$.

The operation $S(\alpha)$, a rotation by α followed by a reflection in the xy plane, will invert the p_z orbital and mix the p_x and p_y orbitals in exactly the same way as $C(\alpha)$. Hence the character for this operation will be $(-1 + 2\cos\alpha)$.

The characters for the different rotation operations of O_h may readily be obtained from these general expressions. The characters for the reflection operation (σ) and of inversion at the centre of symmetry (i) may be treated as operations of the type $S(\alpha)$. Thus $\sigma \equiv S(2\pi)$ and $i \equiv S(\pi)$. Hence

$$\chi(\sigma) = -1 + 2 = +1$$
$$\chi(i) = -3.$$

The set of characters obtained for the basis $\widetilde{p_x p_y p_z}$ is as follows.

E	$8C_3$	$6C_2$	$6C_4$	$6C_2'$	i	$8S_6$	$3\sigma_h$	$6S_4$	$6\sigma_d$
3	0	-1	1	-1	-3	0	1	-1	1

Inspection of the character table for O_h shows that these are the characters of the T_{1u} irreducible representation. The p orbitals on M therefore form a basis for this irreducible representation.

We apply the same procedure to determine the irreducible representations for which the d orbitals of M form a basis. The equations giving the result of a rotation by an angle α of each of the d orbitals in turn are as follows.

$$C(\alpha) \; d_{x^2-y^2} = \cos 2\alpha \; d_{x^2-y^2} + \sin 2\alpha \; d_{xy}$$

$$C(\alpha) \; d_{xy} = -\sin 2\alpha \; d_{x^2-y^2} + \cos 2\alpha \; d_{xy}$$

$$C(\alpha) \; d_{xz} = \cos \alpha \; d_{xz} + \sin \alpha \; d_{yz}$$

$$C(\alpha) \; d_{yz} = -\sin \alpha \; d_{xz} + \cos \alpha \; d_{yz}$$

$$C(\alpha) \; d_{z^2} = d_{z^2} \; .$$

Hence $$\chi[C(\alpha)] = 2 + 2\cos\alpha + 2\cos 2\alpha$$

and $$\chi[S(\alpha)] = 2 - 2\cos\alpha + 2\cos 2\alpha.$$

Using these expressions, the set of characters obtained is

E	$8C_3$	$6C_2$	$6C_4$	$6C_2'$	i	$8S_6$	$3\sigma_h$	$6S_4$	$6\sigma_d$
5	-1	1	-1	1	5	-1	1	-1	1

In this case the characters belong to a reducible representation of O_h. By inspection of the character table of O_h it may readily be seen that the above set of characters is a linear combination of the characters for the irreducible representations E_g and T_{2g}.

The orbitals d_{xy}, d_{yz}, d_{xz}, which all have their lobes directed between four fluorine atoms will evidently form a basis for the T_{2g} representation and the d_{z^2} and $d_{x^2-y^2}$ orbitals will form a basis for the E_g representation.

We now consider the symmetry orbitals which can be formed from the s and p orbitals of the fluorine ligands. The p orbitals on the fluorine atoms are chosen so that they are either directed along the MF axis with their positive lobes pointing towards M or perpendicular to this axis. By convention, the p orbitals directed along the MF axis are labelled σ orbitals and those perpendicular to the MF axis are labelled π orbitals. It is easily verified that no operation of the group O_h will transform a σ orbital into a π orbital and vice versa. The two groups may therefore be treated separately. The s orbitals on the fluorine atoms belong with the σ orbitals. It should be pointed out here that most of the computer programmes for *ab initio* calculations do not

74

allow any choice in the orientation of the p orbitals on the ligand atoms. The p orbitals must in general be oriented so that they are oriented in the same direction as one of the axes of the global coordinate system of the molecule, With this restriction, the p_σ orbital centred on atom B will be the p_x orbital centred on B, but the p_σ orbital on E will be $-p_x$ on E. The p_σ orbital on F will be the p_y orbital on F, but the p_σ orbital on C will be $-p_y$ on C.

The characters of the reducible representation for which the σ orbitals form a basis may be obtained without difficulty, because in general any operation of the O_h group will either leave a given σ orbital unchanged or transform it into another member of the set. The orbital will therefore contribute either 1 or 0 to the character of the reducible representation, as explained in the preceding example. The set of characters obtained for the σ orbitals is

E	$8C_3$	$3C_2$	$6C_4$	$6C_2'$	i	$8S_6$	$3\sigma_h$	$6S_4$	$6\sigma_d$
6	0	2	2	0	0	0	4	0	2

This set of characters may be decomposed into the sum of the characters of the irreducible representations A_{1g}, E_g, and T_{1u}. The σ orbitals therefore form a basis for these irreducible representations.

The exact form of the symmetry orbitals belonging to each of these irreducible representations may be obtained by applying the projection operators which were described in the preceding example. Applying the operator for the A_1 representation to the orbital σ_A, we obtain

$$
\begin{aligned}
P_{A_1}\, \sigma_A = \; & \sigma_A + (2\sigma_A + \sigma_B + \sigma_C + \sigma_E + \sigma_F) + \\
& + (\sigma_A + 2\sigma_D) + (2\sigma_D + \sigma_B + \sigma_C + \sigma_E + \sigma_F) + \\
& + (2\sigma_B + 2\sigma_C + 2\sigma_E + 2\sigma_F) + \sigma_D + \\
& + (2\sigma_D + \sigma_B + \sigma_C + \sigma_E + \sigma_F) + (2\sigma_A + \sigma_D) + \\
& + (2\sigma_A + \sigma_B + \sigma_C + \sigma_E + \sigma_F) + \\
& + (2\sigma_B + 2\sigma_C + 2\sigma_E + 2\sigma_F) \\
= \; & 8(\sigma_A + \sigma_B + \sigma_C + \sigma_D + \sigma_E + \sigma_F).
\end{aligned}
$$

The numerical factor is replaced by a normalizing coefficient by the programme. Similarly, applying the operator for T_{1u} to σ_A we obtain the symmetry orbital

$$\Psi_1\,(T_{1u}) = (\sigma_A - \sigma_B).$$

This is one component of the set of three symmetry orbitals belonging to the representation T_{1u}. The other two components may be obtained by applying the same operator to the orbitals σ_C and σ_E.

$$\Psi_2 \, (T_{1u}) \;=\; (\sigma_B - \sigma_E)$$
$$\Psi_3 \, (T_{1u}) \;=\; (\sigma_C - \sigma_F).$$

We could have obtained these symmetry orbitals without using the projection operators by forming linear combinations of the σ orbitals which have the same symmetry as the p orbitals on the central atom, because these three orbitals also form a basis for the irreducible representation T_{1u}. The combination $(\sigma_A - \sigma_D)$ has the same symmetry as p_z, the combination $(\sigma_C - \sigma_F)$ has the same symmetry as p_y, and the combination $(\sigma_B - \sigma_E)$ has the same symmetry as p_x.

For the E_g symmetry orbitals we have the same problem as we had for the E symmetry orbitals of NH_3. We can project three linearly dependent symmetry orbitals from which we must form two functions which belong to different rows of the degenerate representation. The two functions obtained in this way are

$$\Psi_1 \, (E_g) \;=\; (\sigma_B - \sigma_C + \sigma_E - \sigma_F) \qquad \text{and}$$
$$\Psi_2 \, (E_g) \;=\; (2\sigma_A - \sigma_B - \sigma_C + 2\sigma_D - \sigma_E - \sigma_F).$$

These two symmetry orbitals could equally well have been obtained by forming linear combinations of the σ orbitals which have the same symmetry as the $d_{x^2-y^2}$ and d_{z^2} orbitals on the central atom.

The π orbitals on the fluorine atoms may be further divided into two sets, labelled π and π' respectively, which are shown in the diagram below.

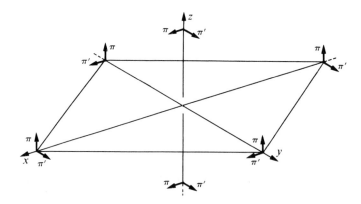

To determine the reducible representation for which the π and π' orbitals form a basis we follow the same procedure that we used for the σ orbitals. This gives the following set of characters:

E	$8C_3$	$3C_2$	$6C_4$	$6C_2'$	i	$8S_6$	$3\sigma_h$	$6S_4$	$6\sigma_d$
12	0	-4	0	0	0	0	0	0	0

This may be decomposed as before to give

$$T_{1g} + T_{2g} + T_{1u} + T_{2u}.$$

With the projection operator for T_{1u} the following functions are obtained.

$$\Psi_1 (T_{1u}) = (\pi'_C + \pi'_F + \pi_A + \pi_D)$$

$$\Psi_2 (T_{1u}) = (\pi'_E + \pi'_B + \pi'_A + \pi'_D)$$

$$\Psi_3 (T_{1u}) = (\pi_E + \pi_B + \pi_C + \pi_F)$$

These functions can also be obtained by forming combinations of the π orbitals which have the same symmetry as the p orbitals on the central atom.

$$\Psi_1 (T_{2g}) = (\pi'_E - \pi'_B + \pi'_C - \pi'_F)$$

$$\Psi_2 (T_{2g}) = (\pi_E - \pi_B + \pi_A - \pi_D)$$

$$\Psi_3 (T_{2g}) = (\pi_C - \pi_F + \pi'_A - \pi'_D)$$

The projection operators give for the symmetry orbitals of symmetry T_{1g} and T_{2u}

$$\Psi_1 (T_{1g}) = (\pi_C - \pi_F - \pi'_A + \pi'_D)$$

$$\Psi_2 (T_{1g}) = (\pi_E - \pi_B - \pi_A + \pi_D)$$

$$\Psi_3 (T_{1g}) = (\pi'_E - \pi'_B - \pi'_C + \pi'_F)$$

$$\Psi_1 (T_{2u}) = (\pi'_C + \pi'_F - \pi_A - \pi_D)$$

$$\Psi_2 (T_{2u}) = (\pi'_E + \pi'_B - \pi'_A - \pi'_D)$$

$$\Psi_3 (T_{2u}) = (\pi_E + \pi_B - \pi_C - \pi_F)$$

There are no orbitals of T_{1g} or T_{2u} symmetry on the central atom and so the alternative method is not available to us for these irreducible representations.

In addition to the symmetry orbitals we also need to know the molecular orbital configuration of the molecule in the ground state or in the particular excited state that we wish to calculate. The inner shells of the metal atom and the 1s and 2s shells of the fluorine atom may be treated separately, as they take only a negligible part in the bonding. The inner s and p shells of the metal atom will have the same order as in the isolated atom and the corresponding m.o. will have a_{1g} or t_{1u} symmetry. The relative position of the orbitals obtained from the 1s and 2s orbitals of the fluorine atoms may be determined by comparing the orbital energies of the metal and the fluorine atom obtained by means of Hartree–Fock calculations on the isolated atoms.

The orbitals that take part in the bonding will be the 3d, 4s, and 4p orbitals on the metal atom and the 2p orbitals on the fluorine atom. The way in which these orbitals interact has been established using qualitative arguments and is usually shown in the form of a diagram similar to that below.

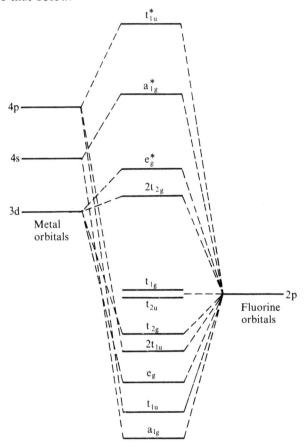

The valence electrons from the metal atom and the fluorine ligands are then allocated to these molecular orbitals, filling each orbital in turn, starting with the orbital of lowest energy. All the orbitals will be completely filled as far as the t_{1g} orbital. The way in which the remaining electrons are distributed between the $2t_{2g}$ and e_g^* orbitals depends on the difference in energy between these orbitals, so we must use spectroscopic evidence to complete the molecular orbital configuration.

11

THE CORRELATION-ENERGY PROBLEM

What has been said so far presents a very rosy picture. *Ab initio*
analytical solutions of the Hartree–Fock equations may be found both
for diatomic and polyatomic molecules. No integrals need be approximated by semi-empirical guesses, and programmes are currently available
to compute wave functions and hence mean values of observables for
molecules of any geometry, containing in principle up to something like
fifty atoms and two hundred electrons both for closed and open shells.

This is indeed a happy situation, but it should not be imagined
that these wave functions are the last word in sophistication or indeed
that they are even highly satisfactory. Providing enough care is taken
in choosing a good basis set and enough computer time is available to
use large bases, the results can approach the Hartree–Fock limit; that
is the best possible solution to the Hartree–Fock equations, equal to
that which would be obtained were the equations to be solved numerically
or with an infinite basis set.

We can in principle get the best possible answer within the limits
of the Hartree–Fock equations. Herein lies the problem. The Hartree–
Fock equations only approximately represent the reality of the molecular
situation and are subject to some severe limitations.

11.1. The relativistic energy

No mention is made of relativistic corrections in the derivation
of the Hartree–Fock equations. The virial theorem tells us that inner-
shell electrons with the highest potential energy will have the greatest
kinetic energy, so that relativistic effects may become important for
these electrons. This is particularly true of the inner electrons of
heavy atoms or molecules, e.g. for Mg the relativistic energy of a 1s
electron is thought to be about 0·2 a.u. and that for a 2s electron
0·03 a.u. (1 a.u. = 27·21 eV).

Relativistic energies are certainly important contributors to the
total energy of molecules containing heavy atoms and are hard to
estimate. Fortunately, however, quite frequently it is not absolute
energies that are of interest, but rather differences between energy

levels, and since the inner-shell electrons are normally unchanged when going from one electronic energy level to another the problem of relativistic energy may frequently be ignored.

11.2. The correlation energy

Far more serious is the correlation energy. This is defined as follows:

$$E_{\text{total}\atop\text{experimental}} = E_{\text{Hartree-Fock}} + E_{\text{relativistic}} + E_{\text{correlation}}$$

It is the residual error in Hartree—Fock calculations after taking account of the non-relativistic nature of the solutions.

The physical origin of this error is not hard to find although this is not much help in overcoming the difficulty. In the Hartree—Fock equation the inter-electronic interaction is represented by coulomb and exchange terms; each electron having a direct interaction with the averaged-out charge of all the others obtained by squaring the one-electron wave functions, but an exchange interaction only with electrons of the same spin.

In reality an electron in an atom will have instantaneous interactions with all the other electrons which will not be the same as the average interaction included in the SCF procedure. This is the origin of the correlation-energy error, which is partly accounted for in the case of electrons of the same spin by the exchange terms in the Hartree—Fock equations.

By systematically examining the magnitudes of $E_{\text{correlation}}$ for series of atoms and ions, Sinanoglu and Clementi have shown that the effects can be largely accounted for as sums of pair effects between all pairs of electrons in the atom. Largely speaking, the dominant pair effects come from effects between pairs of electrons in the same spatial orbital, between which, of course, there are no exchange terms.

Tables[24] of estimates of the magnitude of these pair correlation effects have been published, but the picture is complicated, as it has also been demonstrated that additional contributions to correlation energy come from near-degeneracy effects.[25] This latter contribution can be considered as configuration interaction with near-by states, but again this is hard to treat precisely.

Crude attempts can be made to surmount the correlation energy problem in molecules by using an atomic population analysis which assigns on a percentage basis the constituent a.o. in an m.o. and then using tables of atomic correlation energies. This is not a very satisfactory procedure, but there is often no alternative.

Some promising theoretical work aims at direct calculation of the correlation energy in atoms, but for molecules it would be true to say that the correlation-energy problem remains one of the major difficulties in theoretical chemistry.

Happily, despite this serious error there are some important applications of *ab initio* wave functions for which it is unimportant. The following diagram indicates a typical computed potential energy curve of a diatomic molecule compared with the true experimental curve.

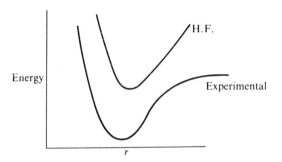

Two things should be noticed. Firstly the Hartree—Fock curve lies above the experimental, but it is closest to being parallel at the minimum. Secondly the computed curve is narrower at the bottom of the well. As a result the computed r_e values are normally close to observed results (frequently some hundredths of an Ångström unit too small) and the vibration frequencies are too large. The same situation occurs in the multidimensional potential surfaces of polyatomic molecules.

The fact that the computed curves are most realistic at the minimum is a happy result since it means that in polyatomic calculations the equilibrium geometries of states are normally rather satisfactorily predicted.

In cases where one is computing differences in energies, and where no difference in correlation energy between the two situations is to be expected, *ab initio* computations can give extremely good answers. For instance the repulsive portion of the potential between two rare-gas atoms can be found rather well by calculating the variation of the energy of the He—He system with distance. In the interaction, no new pairings of electrons are introduced, so the correlation energy error will be constant with r giving a calculated potential parallelling the true one. Likewise if energies of the ethane molecule are computed at various staggered and eclipsed positions the differences again agree remarkably well with experimental barriers, since there are no differences in pairings of electrons in the various positions nor are the near-degeneracy effects likely to be important.

An *ab initio* study [26] of the reaction $NH_3 + HCl \rightleftharpoons NH_4Cl$ has been carried out. On the basis of the calculations it was predicted that NH_4Cl should be stable enough to exist in the gas phase though dissociation is almost complete at $400°C$. This prediction has recently been confirmed by the observation of deuterated ammonium chloride molecules in a mass spectrometer. [27] The observed dissociation energy is in excellent agreement with the predicted energy. Again the success of the computation can be attributed to the fact that no new electron pairs are produced when the hydrogen bond is formed.

12

RELATIONSHIP BETWEEN *AB INITIO* AND SEMI-EMPIRICAL M.O. CALCULATIONS

Not even the wildest optimist would expect *ab initio* wave functions for large polyatomic molecules to be routinely computed in the near future. Nonetheless the Hartree−Fock equations

$$\left\{ H^{N} + \sum_{j} J_{j} - \sum_{j}' K_{j} \right\} \phi_{i} = \epsilon_{i}^{SCF} \phi_{i}$$

are applicable no matter how many electrons there are in the molecule.

For large polyatomic molecules a variety of approximations to these equations and their solution have been made, and it is the purpose of this chapter to indicate the relationships between the approximate and *ab initio* methods. We consider first those approximations that can be applied to any type of molecule and then the more commonly used approximation, which are restricted to π-electron systems and have been so successful.

12.1. The SCF equations

All the commonly used approximate m.o. methods start with the Roothaan form of the SCF equations. Summarizing what has gone before, we express each m.o. ϕ_{i} as a linear combination of a.o.s.

$$\phi_{i} = \sum_{n} c_{in} \chi_{n},$$

and the Roothaan equations are then

$$\sum_{n} H_{mn}^{SCF} c_{in} = \sum_{n} S_{mn} c_{in} \epsilon_{i}$$

where

$$H_{mn}^{SCF} = H_{mn}^{N} + G_{mn}$$

and

$$H_{mn}^{N} = \int \phi_{m}^{*} \left[-\frac{1}{2} \nabla^{2} - \sum_{A} \frac{Z_{A}}{r_{A}} \right] \phi_{n} \, d\tau$$

$$G_{mn} = \sum_{ls} \left[P_{ls} (mn|ls) - \frac{1}{2} (ms|ln) \right]$$

$$(mn|ls) = \int \int \phi_m^* (1) \phi_n (1) \frac{1}{r_{12}} \phi_l^* (2) \phi_s (2) \, d\tau_1 \, d\tau_2$$

$$S_{mn} = \int \phi_m^* \phi_n \, d\tau$$

$$P_{ls} = 2 \sum_i^{occ} c_{il}^* c_{is}$$

The various approximate methods make increasingly drastic assumptions about the integrals which appear in the above expressions, some being set equal to zero and others replaced by semi-empirical estimates.

12.2. The valence-electron approximation

A rather reasonable approximation is to consider all 1s shells on the atoms constituting a molecule to be part of an unpolarizable core and to treat only the valence electrons.

This will alter the expression given above for H_{mn}^N to

$$H_{mn} = \int \phi_m^* \left[-\frac{1}{2} \nabla^2 - \sum_A V_A(r) \right] \phi_n \, d\tau$$

where $\sum_A V_A(r)$ is the electrostatic field of the core written as a sum of the potentials $V_A(r)$ for the various atoms in the molecule.

E_{total} will now be equal to $E_{electronic} + E_{core}$

$$= \frac{1}{2} \sum_{mn} P_{mn} (H_{mn}^N + G_{mn}) + \sum_A \sum_B \frac{Z_A' Z_B'}{R_{AB}}$$

where Z_A' is the *core* charge on A.

In making further approximations care must be taken to ensure that the results are invariant to simple transformations of the basis set, such as rotations of axes or replacing simple s and p orbitals by hybrids. This has been considered in detail by Pople, Santry, and Segal.[28]

12.3. Neglect of diatomic differential overlap (NDDO)

In this rather high-level approximation to the Roothaan equations, as well as making the valence-shell approximation the X_m are treated as if they form an orthonormal set, so that S_{mn} is zero unless $m = n$, in which case $S_{mm} = 1$. The coefficients c_{im} then obey the relationship

$$\sum_m c_{im}\, c_{jm} = \delta_{ij} \quad \text{and thus} \quad \sum_m P_{mm} = 2N.$$

In addition differential overlap in two-electron integrals is neglected. This means that the overlapping charge densities of basis orbitals on different atoms are neglected; $(mn \mid ls)$ is zero unless m and n belong to the same atom A and l and s to B.

The remaining integrals may be calculated from a small basis set of a.o. or chosen empirically.

A method known as MINDO due to Dewar is rather similar to this and details are given in his recent book. [33]

12.4. Complete neglect of differential overlap (CNDO)

This increasingly popular approximation goes one stage further than the NDDO method. All two-electron integrals which depend on the overlapping of charge densities of different atomic basis orbitals are neglected,

i.e. $(mn \mid ls) = 0$ unless $m = n$ and $l = s$ in which case the integral is often written $\gamma_{ml} = (mm \mid ll)$. This is a fairly drastic approximation since some one-centre integrals are ignored, e.g. $(2s_A\, 2p_{xA} \mid 2s_A\, 2p_{xA})$.

The electron-interaction integrals γ_{mn} are assumed to be dependent only on the atoms to which X_m and X_n belong and not to the particular type of a.o. As a result one is left with a single set of integrals γ_{AB} interpretable as an average repulsion between an electron in a valence shell on atom A and another valence electron in an a.o. on B.

The matrix elements now become

$$H^{SCF}_{mm} = H^{N}_{mm} + \tfrac{1}{2} P_{mm}\, \gamma_{mm} + \sum_{s(\neq m)} P_{ss}\, \gamma_{ms}$$

$$H^{SCF}_{mn} = H^{N}_{mn} - \tfrac{1}{2} P_{mn}\, \gamma_{mn}$$

or even more simply

$$H^{SCF}_{mm} = H^{N}_{mm} - \tfrac{1}{2} P_{mm}\, \gamma_{AA} + \sum_{A} P_{AA}\, \gamma_{AA} + \sum_{B(\neq A)} P_{BB}\, \gamma_{AB}$$

where m belongs to atom A and $P_{BB} = \sum_n P_{nn}$

The core matrix elements can be even further developed

$$H^{N}_{mm} = \langle m \mid -\tfrac{1}{2}\nabla^2 - V_A \mid m \rangle - \sum_{B(\neq A)} \langle m \mid V_B \mid m \rangle$$

Here the first term measures the energy of the atomic orbital m on A and can either be computed using suitably chosen a.o.s or estimated semi-empirically. The second term gives the interaction of an electron in X_m with the cores of atoms B.

Similarly
$$H_{mn}^N = U_{mn} - \sum_{B(\neq A)} <m|\, V_B\, |n>.$$

Again U_{mn} is the one-electron matrix element involving the local core Hamiltonian and may be zero by symmetry if, for example m and n are respectively s and p functions. The final term represents the interaction of the distribution $X_m X_n$ with the cores of other atoms. Such integrals $<m|\, V_B\, |n>$, when m and n belong to the same atom A, are put equal to zero if $m \neq n$ and V_{AB} if $m = n$ written; then

$$H_{mm}^N = U_{mm} - \sum_{B(\neq A)} V_{AB} \qquad (m \text{ on atom } A)$$

and
$$H_{mn}^N = 0 \qquad (m \neq n \text{ but both on atom } A).$$

When m and n are on different atoms, H_{mn}^N is written as β_{mn} and given by $H_{mn}^N = \beta_{mn} = \beta_{AB}^0 S_{mn}$, with β_{AB}^0 as a parameter depending only on atoms A and B.

Now finally the LCAO–SCF equations can be solved, most of the atomic integrals having been eliminated or set equal to parameters.

This type of approximation, while not giving good energies or spectral information, does predict correct conformations of molecules, rather paralleling the behaviour of *ab initio* calculations which has been mentioned before. Bond lengths and vibration frequencies are predicted very poorly.

A further improvement called CNDO/2 which differs from the above in some small details is now widely used in many laboratories. [29]

12.5. Use of an undefined Hamiltonian (extended Hückel theory)

Very daring in their approximations but useful, almost powerful, in practice are the extended Hückel and similar theories. [30] These again start with the Roothaan equations but leave the Hamiltonian undefined.

$$H \sum_n c_{in}\, X_n = \epsilon_i \sum_n c_{in}\, X_n$$

and the coefficients c_{in} and the orbital energies ϵ_i are obtained by solving the secular determinant

$$\det |H_{nm} - \epsilon\, S_{nm}| = 0$$

The explicit form of H_{mn} is never sought and nearly all integrals are represented by empirical parameters. The complete secular determinant is treated, all interactions are accounted for and the off-diagonal terms are retained.

Matrix elements H_{mm} are taken as measures of the electron attracting power of particular atoms, so that for example in a calculation on a saturated hydrocarbon values for the carbon sp^3 valence state might be

$$H_{mm} \text{ (C2p)} = -11 \cdot 4 \text{ eV.}$$

$$H_{mm} \text{ (C2s)} = -21 \cdot 4 \text{ eV.}$$

The H_{mn} are often approximated as

$$H_{mn} = 0 \cdot 5 \, K \, (H_{mm} + H_{nn}) \, S_{mn} \,;$$

K is a parameter and the S_{mn} are computed from analytic a.o.s centred on the appropriate atoms.

The matrix elements being fixed, a simple diagonalization of the determinant yields the ϵ_i and hence the coefficients c_{in} of the m.o.s. There is, of course, no self-consistency as the resulting orbitals are not used in computing the matrix elements.

The total Hückel energy is taken as $\Sigma 2\epsilon_i$ for a closed-shell molecule and the coefficients can be used to determine approximate charge distributions in the molecule.

Despite the very crude nature of the method, rather good predictions of geometries can be obtained, but since the electron interaction is not specifically considered there is no such concept as exchange, and singlet and triplet states are not distinguished.

12.6. The pi-electron approximation

There are a whole series of approximation m.o. methods which roughly follow the hierarchy of complexity so far discussed in this chapter but which contain an additional assumption — the pi-electron approximation, which has been so successful for conjugated systems.

Physically this amounts to assuming that we can consider a series of states of a conjugated molecule as having a constant sigma-electron framework and that variations which exist are only due to the pi electrons. Thus the complete wave function is

$$\Psi = \Sigma\Pi,$$

where Σ and Π are proper antisymmetric wave functions describing the appropriate sets of electrons.

The Hamiltonians can be written

$$H = H_\sigma^0 + H_\pi^0 + H_{\sigma\pi},$$

where H_σ^0 and H_π^0 are of the usual Hartree—Fock form but refer only to σ and π electrons respectively,

$$H_{\pi\sigma} = \sum_i^{n\sigma} \sum_j^{n\pi} \frac{1}{r_{ij}}$$

The pi-electron Hamiltonian is defined as

$$H^{\pi} = H^0_{\pi} + H_{\pi\sigma}$$

and $H_{\pi\sigma}$ included in the core portion,

i.e.
$$H^{\pi} = \sum_i^{n\pi} h^{\text{core}} + \sum_{i<j}^{n\pi} \frac{1}{r_{ij}}$$

The Hartree–Fock operator is then

$$H = h^{\text{core}} + \sum_j^{n\pi} (2J_j - K_j) = h^{\text{core}} + G_{\pi}.$$

Solutions to these equations and approximations thereto make up the wealth of π-electron calculations which have been published. These have been dealt with in a number of books[4,31,32,] which give full details of the approximations, but we give here brief outlines for completeness.

(a) The Goeppert-Meyer–Sklar method

This method works within the π-electron approximation but makes very few further assumptions. The m.o. are expanded in a basis of $2p_z$ a.o.s, assumed to be orthogonal. The J and K integrals are calculated completely except that three- and four-centre integrals are neglected, but h^{core} is simplified by approximation. Sigma–pi effects are neglected as are the effects of hydrogen nuclei. Then

$$h^{\text{core}} = -\tfrac{1}{2} \nabla_i^2 + \sum_s V_{si}$$

and we assume

$$h^{\text{core}} \phi_i = W_{2p} \phi_i \, ,$$

and W_{2p} is regarded as the ionization potential of carbon in an sp^2 valence state.

(b) The Pariser – Parr – Pople method

There are several possible modifications of this particular scheme, but most are at about the same level of sophistication. The starting point is the Roothaan equation in the form

$$\{h^{\text{core}} + 2J_i - K_i\} \sum_n c_{in} \chi_n = \epsilon_i \sum_n c_{in} \chi_n$$

The CNDO approximation is made, i.e. $S_{mn} = \delta_{mn}$ and any integral of the form

$$\int \int \chi_m^* (1) \, \chi_n (1) \, \frac{1}{r_{12}} \, \chi_l^* (2) \, \chi_s (2) \, d\tau_1 \, d\tau_2$$

is set equal to
$$\delta_{mn} \, \delta_{ls} \, \gamma_{ml}$$

and
$$\gamma_{ml} = \int \int \chi_m^2 (1) \, \frac{1}{r_{12}} \, \chi_l^2 (2) \, d\tau_{12}$$

is taken as an empirical parameter.

For the core integrals a type of Goeppert – Meyer–Sklar approximation is made, $h_{mn}^{core} = 0$ for $m \neq n$ unless m and n are on neighbouring atoms, when $h_{mn}^{core} = \beta_{mn}$, another empirical parameter.

$$h_{mm}^{core} = \int \chi_m \left[-\frac{1}{2} \nabla^2 - V_m \right] \chi_m \, d\tau + \sum_{m(\neq n)} \int \chi_m V_n \chi_m \, d\tau,$$

where V_m and V_n are the potentials due to atoms m and n. The first of these terms is replaced by an empirical parameter I_m, a modified ionisation potential, and the second term approximated by $-Z_n \gamma_{mn}$, representing the electron nuclear attraction with other atoms.

Despite the great simplifications made the P.P.P. method still requires the use of a computer, but fairly large systems may be treated.

(c) The Hückel method

The very well-known and successful Hückel method is the most drastic of all the approximations. It is within the pi-approximation and furthermore assumes an effective Hamiltonian H that does not treat interelectron repulsions specifically.

$$H \sum_n c_{in} \, \chi_n = \epsilon_i \sum_n c_{in} \, \chi_n$$

giving secular equations

$$\sum_n c_{in} \, (H_{mn} - \epsilon S_{mn}) = 0,$$

which only have non-trivial roots when

$$\det | H_{mn} - \epsilon S_{mn} | = 0.$$

To solve this determinant very far-reaching assumptions are made about matrix elements.

$H_{mm} = \alpha$, an empirical parameter,

$H_{mn} = \beta$, (if m and n are neighbours) an undefined parameter,

$H_{mn} = 0$, if m and n are not neighbours,

$S_{mn} = \delta_{mn}$.

Energies are normally quoted in β units.

12.7. Summary

No attempt has been made to treat the various approximations to Hartree–Fock solutions fully. Only the relationships between the methods have been stressed and the position of each in the hierarchy of complexity indicated. Each method could be, and in some cases has been, the subject of a lengthy monograph, and it is to such works that the reader must go for much essential detail.

13
CONCLUSIONS

In this short book no attempt has been made at mathematical rigour.
Indeed all the theory, such as group theory, which is covered in a
number of quantum mechanical texts, has been ignored, and the reader
is left to go to these for such amplification as he feels is necessary.
On the other hand, a serious attempt has been made to enable a chemist
with a normal undergraduate training in elementary wave mechanics to
compute *ab initio* wave functions for problems that interest him, using
one of the published computer programmes, and furthermore to utilize
the results.

The ready availability of these massive *ab initio* wave function
programmes is due to the generosity of the men who have built and
developed them. They provide a striking example of cooperation, which
experimentalists might well envy. An experimental worker is normally
loath to allow other scientists, whom he may regard as competitors, to
reap the benefits of his work in developing apparatus. The writers of
programmes have not only been ready to furnish copies of their 'apparatus'
to interested fellow scientists but even provide a service of documentation
which includes detailed instructions and test input and output.

This organization is the Quantum Chemistry Program Exchange
(Q.C.P.E.) organized from Indiana University. Their catalogue now
contains over 120 ready-built programmes, several of which are highly
sophisticated *ab initio* types capable of treating even large polyatomic
systems.

The public-spiritedness of these programme builders is the reason
why a monograph like this is necessary. The big programmes are
available to the people who have the interesting problems to use for
themselves. It is too much for the organic chemist to expect theoreticians to do routine computations for him as a service. He should
be prepared to run the computations for himself, just as he now does
his own Hückel calculations.

The calculation of wave functions and energies has been stressed
here, since this is the area where most work has been done. Energies
of possible configurations of a molecule may be calculated to answer
questions about the nature of the ground state of a molecule which may

have a profound effect on the thermodynamic properties by changing the electronic partition function. Similarly the nature of excited states can be clarified by direct *ab initio* calculations.

Other observables can be computed from the wave functions by using the appropriate operator. In particular, dipole moments and magnetic hyperfine constants have received attention and so, more recently, have spin—orbit coupling constants. The computed values are often in very close agreement with the experimentally determined figures, but, more important, they may enable a sign to be given to a quantity that is only ambiguously given by the experiment.

The wave function itself is very useful since its square gives a charge density which may be computed at any point in space. The charge densities at nuclei are obviously important in interpreting hyperfine interactions in resonance spectroscopy, but, furthermore, charge densities at a distance from molecules are also interesting. In particular, one can see how the wave functions of adjacent molecules in a crystal or polymer might overlap to allow tunnelling of electrons between sites.

Charge densities may be sufficient to explain reaction kinetic behaviour, but *ab initio* calculations can do far more than just provide static indices of charge distributions. Cross-sections through potential surfaces for some reactions have already been calculated.

As computers get bigger and faster the chemist is going to want more accurate wave functions than the approximate methods can yield. Particularly for excited states where experimental data are hard to provide, the wave functions are certain to become very important.

If a property can be measured then the experimental scientist will normally prefer a measured value to a computed one, but for excited states of molecules, which are so important in photochemical and astrophysical work, there may be no choice. The only way to find a charge distribution or the probable geometry may be by computation and in which case the more realistic the wave function the better.

In 1960 about ten papers on *ab initio* computations of wave functions were published. In 1967 the number was several hundred. A complete bibliography of *ab initio* molecular wave functions up to 1969 has been published.[34] The trend is certain to continue, and to the extent that the technique is certain to become commonplace amongst chemists of every variety.

APPENDIX 1

Examples illustrating the taking of matrix elements between Slater determinants

The most tricky rule to use is the one involving matrix elements between Slater determinants with two spin orbitals different. Let us illustrate the use of this by means of a real example which would form part of a calculation of configuration interaction on the ground state of nitrogen.

The ground state wave function is a single determinant.

$$\Psi_0 = |\, 1\sigma_g^2 \, 1\sigma_u^2 \, 2\sigma_g^2 \, 2\sigma_u^2 \, 3\sigma_g^2 \, 1\pi_u^4 \,|.$$

If we now want to consider the interaction of those states of similar ($^1\Sigma_g^+$) symmetry obtained by promoting two π_u electrons to π_g orbitals we will firstly have to consider the following determinants (writing only the open-shell part).

$$\psi_1 = |\, \ldots \, 1\pi_u^+ \, 1\pi_u^- \, 1\overline{\pi_g^+} \, 1\overline{\pi_g^-} \,|$$

$$\psi_2 = |\, \ldots \, 1\pi_u^+ \, 1\overline{\pi_u^-} \, 1\pi_g^+ \, 1\overline{\pi_g^-} \,|$$

$$\psi_3 = |\, \ldots \, 1\overline{\pi_u^+} \, 1\pi_u^- \, 1\pi_g^+ \, 1\overline{\pi_g^-} \,|$$

$$\psi_4 = |\, \ldots \, 1\pi_u^+ \, 1\overline{\pi_u^-} \, 1\overline{\pi_g^+} \, 1\pi_g^- \,|$$

$$\psi_5 = |\, \ldots \, 1\overline{\pi_u^+} \, 1\pi_u^- \, 1\overline{\pi_g^+} \, 1\pi_g^- \,|$$

$$\psi_6 = |\, \ldots \, 1\overline{\pi_u^+} \, 1\overline{\pi_u^-} \, 1\pi_g^+ \, 1\pi_g^- \,|$$

Combinations of these will give rise to two singlet states as may be seen from the table given in Appendix 2 for the case of singlet functions from four orbitals, a, b, c, d, outside the closed-shell part. In fact from the table we can see that these functions will be

$$\Psi_A = \tfrac{1}{2}(\psi_1 + \psi_6 - \psi_3 - \psi_4)$$

and $\quad \Psi_B = \dfrac{1}{\sqrt{3}}\left(\psi_2 - \tfrac{1}{2}\psi_1 + \psi_5 - \tfrac{1}{2}\psi_6 - \tfrac{1}{2}\psi_3 - \tfrac{1}{2}\psi_4\right)$

(Note that the order the functions are written $\psi_1 - \psi_6$ above does not not correspond to the order which the functions, $A - F$ involving a, b, c, d are written in the table, and care must always be exercised when using

93

the tables of Appendix 2 to ensure that a list of functions such as $\psi_1 - \psi_6$ is matched exactly with ψ_A, etc. in the tables.)

There will be another $^1\Sigma_g^+$ state from a rather similar configuration which is

$$\Psi_c = \frac{1}{\sqrt{2}} (\psi_7 - \psi_8),$$

where

$$\psi_7 = | \dots 1\pi_u^+ \, 1\overline{\pi_u^+} \, 1\pi_g^- \, 1\overline{\pi_g^-} |$$

and

$$\psi_8 = | \dots 1\pi_u^- \, 1\overline{\pi_u^-} \, 1\pi_g^+ \, 1\overline{\pi_g^+} |$$

Ψ_A, Ψ_B, and Ψ_C are Σ^+ states, since if one reflected ($\pi^+ \leftrightarrow \pi^-$) they they remain the same.

In a C.I. calculation matrix elements such as

$$<\Psi_A | H | \Psi_A >, \quad <\Psi_A | H | \Psi_B >, \quad <\Psi_B | H | \Psi_B >,$$

etc. would be needed. These could be simplified to an extent by using the method of taking matrix elements between projected wave functions using the auxiliary coefficients in the tables of Appendix 2 as described in Chapter 4. However, when this has been done there will still remain a large number of terms involving matrix elements between the single determinants,

$$<\Psi_B | H | \Psi_A > = \sqrt{\left(\frac{3}{4}\right)} \, [H_{21} + H_{26} - H_{23} - H_{24}]$$

What now follows is a list of applications of Slater's rule to these integrals $<\psi_1 | H | \psi_2>$, etc. which it is hoped will help those who are unfamiliar with this process to practise and to check their own application of the rule.

Remember that before applying the rule the two determinants must be altered by exchanging columns to achieve maximum coincidence, each change of column causing a single change in sign. Furthermore any integral of the type

$$\zeta_{\bar{c}d}^{a\bar{b}}$$

or in another shorthand form $(a\bar{b} | \bar{c}d)$, will be zero since α and β functions of a particular electron are orthogonal. As a result there will only be two terms in the case of two orbitals differing if all the electrons involved are of the same spin.

These rather confusing statements are probably best understood by checking the following examples. For the first few a lot of detail will be given but then just the answers so as to provide some incentive for the reader to test the writers' algebra.

$$\langle \psi_1 | H | \psi_0 \rangle = H_{10}$$

$$| \ldots 1\pi_u^+ \overline{1\pi_u^+} \, 1\pi_u^- \overline{1\pi_u^-} |$$

$$| \ldots 1\pi_u^+ \overline{1\pi_u^-} \, \overline{1\pi_g^+} \, \overline{1\pi_g^-} |$$

One change gives maximum coincidence, then applying the rule

$$H_{10} = -\zeta \frac{\overline{1\pi_u^+} \, \overline{1\pi_g^+}}{1\pi_u^- \, 1\pi_g^-} + \zeta \frac{\overline{1\pi_u^+} \, \overline{1\pi_g^-}}{1\pi_u^- \, 1\pi_g^+} = -K^0_{1\pi_u \, 1\pi_g} + K^2_{1\pi_u \, 1\pi_g}$$

$$H_{20}$$

$$| \ldots 1\pi_u^+ \overline{1\pi_u^+} \, 1\pi_u^- \overline{1\pi_u^-} |$$

$$| \ldots 1\pi_u^+ \overline{1\pi_u^-} \, 1\pi_g^+ \overline{1\pi_g^-} |$$

Two changes are required for maximum coincidence, thus there is no change of sign and

$$H_{20} = \zeta \frac{\overline{1\pi_u^+} 1\pi_g^+}{1\pi_u^- \overline{1\pi_g^-}} - \zeta \frac{\overline{1\pi_u^+} \overline{1\pi_g}}{1\pi_u^- 1\pi_g^+}$$

$$= 0 - K^2_{1\pi_u \, 1\pi_g}$$

$$H_{30}$$

$$| \ldots 1\pi_u^+ \overline{1\pi_u^+} \, 1\pi_u^- \overline{1\pi_u^-} |$$

$$| \ldots \overline{1\pi_u^+} \, 1\pi_u^- \, 1\pi_g^+ \overline{1\pi_g^-} |$$

Two changes required for maximum coincidence.

$$H_{30} = \zeta \frac{1\pi_u^+ 1\pi_g^+}{1\overline{\pi_u^-} \, 1\overline{\pi_g^-}} - \zeta \frac{1\pi_u^+ \overline{1\pi_g^-}}{1\overline{\pi_u^-} \, 1\pi_g^+}$$

$$= K^0_{1\pi_u \, 1\pi_g}$$

$$H_{40} = \zeta \frac{\overline{1\pi_u^+} \, \overline{1\pi_g^+}}{1\pi_u^- \, 1\pi_g^-} - 0$$

$$= K^0_{1\pi_u \, 1\pi_g}$$

$$H_{50} = \zeta \frac{1\pi_u^+ \overline{1\pi_g^+}}{1\overline{\pi_u^-} \, 1\pi_g^-} - \zeta \frac{1\pi_u^+ \overline{1\pi_g^-}}{1\overline{\pi_u^-} \, 1\pi_g^+} = -K^2_{1\pi_u \, 1\pi_g}$$

$$H_{60} = -K^0_{1\pi_u \, 1\pi_g} + K^2_{1\pi_u \, 1\pi_g}$$

$$H_{70} = K^0_{1\pi_u \, 1\pi_g}$$

$$H_{80} = K^0_{1\pi_u 1\pi_g}$$

$$H_{12} = -K^2_{1\pi_u 1\pi_g}$$

$$H_{13} = -K^0_{1\pi_u 1\pi_g}$$

$$H_{14} = -K^0_{1\pi_u 1\pi_g}$$

$$H_{15} = -K^2_{1\pi_u 1\pi_g}$$

$$H_{16} = 0 \quad (\psi_1 \text{ and } \psi_6 \text{ differ by more than 2 spin orbitals})$$

$$H_{23} = -J^2_{1\pi_u 1\pi_u}$$

$$H_{24} = -J^2_{1\pi_g 1\pi_g}$$

$$H_{25} = 0$$

$$H_{26} = -K^2_{1\pi_u 1\pi_g}$$

$$H_{34} = 0$$

$$H_{35} = -J^2_{1\pi_u 1\pi_u}$$

$$H_{36} = K^0_{1\pi_u 1\pi_g}$$

$$H_{45} = -J^2_{1\pi_u 1\pi_u}$$

$$H_{56} = -K^0_{1\pi_u 1\pi_g}$$

$$H_{17} = -K^0_{1\pi_u 1\pi_g}$$

$$H_{27} = J^2_{1\pi_u 1\pi_g}$$

$$H_{18} = -K^0_{1\pi_u 1\pi_g}$$

APPENDIX 2

Projected spin functions

In this appendix we give the combinations of open-shell determinants which are proper spin functions. The tables may be used to reduce the number of terms in matrix element expressions as described in Chapter 7.

1. One open-shell electron

We only need to consider the open-shell part of a configuration since the closed-shell part will be totally symmetric.

In this simplest case we only have one possible function $= | \dots \phi_i^{\alpha} |$ or the exactly equivalent $| \dots \phi_i^{\beta} |$, e.g. $= | 1\sigma^2 2\sigma |$. This can only give a doublet state wave function.

2. Two open-shell electrons

(a) Triplet State

The only function will be $| \dots \phi_j^a \phi_k^a |$, which we will write as

$$A = \alpha \alpha$$
$$^3\Psi = A$$

In every case we only need to consider the wave function of one component of a degenerate state, e.g. a triplet and choose here the one with the maximum m_s value since this is the most convenient to work with.

(b) Singlet State

Here we have a resultant spin and this can have two possible determinants :

$$A = \alpha \beta$$
$$B = \beta \alpha.$$

Now we can give the simplest example of the Nesbet projected wave function.

	A	B	k_μ
$^1\Psi$	1*	−1	2

This is interpreted as follows

$$^1\Psi = \frac{1}{\sqrt{2}} [\psi_A - \psi_B]$$

$$<{}^1\Psi | H | {}^1\Psi> = H_{AA} - H_{AB}$$

Note that using the projected spin functions we only have two matrix elements instead of four.

3. Three open-shell electrons

(a) Quartet State

$$A = \alpha\,\alpha\,\alpha$$

	A	k_μ
$^4\Psi$	1*	1

(b) Doublet States

$$A = \alpha\,\alpha\,\beta$$
$$B = \alpha\,\beta\,\alpha$$
$$C = \beta\,\alpha\,\alpha$$

	A	B	C	k_μ
$^2\Psi_1$	$\frac{1}{2}$	1*	-1	2
$^2\Psi_1$	1*	$-\frac{1}{2}$	$-\frac{1}{2}$	$\frac{3}{2}$

4. Four open-shell electrons

(a) Quintet State

$$A = \alpha\,\alpha\,\alpha\,\alpha$$

	A	k_μ
$^4\Psi$	1*	1

(b) Triplet States

$$A = \alpha\,\alpha\,\alpha\,\beta$$
$$B = \alpha\,\alpha\,\beta\,\alpha$$
$$C = \alpha\,\beta\,\alpha\,\alpha$$
$$D = \beta\,\alpha\,\alpha\,\alpha$$

98

	A	B	C	D	k_μ
$^3\Psi_1$	$\frac{1}{2}$	$\frac{1}{2}$	$1*$	-1	2
$^3\Psi_2$	$\frac{1}{2}$	$1*$	$-\frac{1}{2}$	$-\frac{1}{2}$	$\frac{3}{2}$
$^3\Psi_3$	$1*$	$-\frac{1}{3}$	$-\frac{1}{3}$	$-\frac{1}{3}$	4

(c) **Singlet States**

$$A = \alpha\,\beta\,\alpha\,\beta$$
$$B = \alpha\,\alpha\,\beta\,\beta$$
$$C = \beta\,\alpha\,\beta\,\alpha$$
$$D = \beta\,\beta\,\alpha\,\alpha$$
$$E = \beta\,\alpha\,\alpha\,\beta$$
$$F = \alpha\,\beta\,\beta\,\alpha$$

	A	B	C	D	E	F	k_μ
$^1\Psi_1$	$\frac{1}{2}$	$1*$	0	1	-1	-1	4
$^1\Psi_2$	$1*$	$-\frac{1}{2}$	1	$-\frac{1}{2}$	$-\frac{1}{2}$	$-\frac{1}{2}$	3

5. **Five open-shell electrons**

(a) **Sextuplet State**

$$A = \alpha\,\alpha\,\alpha\,\alpha\,\alpha$$

	A	k_μ
$^5\Psi$	1	1

(b) **Quartet States**

$$A = \alpha\,\alpha\,\alpha\,\alpha\,\beta$$
$$B = \alpha\,\alpha\,\alpha\,\beta\,\alpha$$
$$C = \alpha\,\alpha\,\beta\,\alpha\,\alpha$$
$$D = \alpha\,\beta\,\alpha\,\alpha\,\alpha$$
$$E = \beta\,\alpha\,\alpha\,\alpha\,\alpha$$

	A	B	C	D	E	k_μ
$^4\Psi_1$	$\frac{1}{2}$	$\frac{1}{2}$	$\frac{1}{2}$	$1*$	-1	2
$^4\Psi_2$	$\frac{1}{3}$	$\frac{1}{3}$	$1*$	$-\frac{1}{2}$	$-\frac{1}{2}$	$\frac{3}{2}$
$^4\Psi_3$	$\frac{1}{4}$	$1*$	$-\frac{1}{3}$	$-\frac{1}{3}$	$-\frac{1}{3}$	$\frac{4}{3}$
$^4\Psi_4$	$1*$	$-\frac{1}{4}$	$-\frac{1}{4}$	$-\frac{1}{4}$	$-\frac{1}{4}$	$\frac{5}{4}$

(c) Doublet States

$$A = \alpha\,\beta\,\alpha\,\beta\,\alpha$$
$$B = \alpha\,\beta\,\beta\,\alpha\,\alpha$$
$$C = \alpha\,\alpha\,\beta\,\beta\,\alpha$$
$$D = \beta\,\beta\,\alpha\,\alpha\,\alpha$$
$$E = \beta\,\alpha\,\beta\,\alpha\,\alpha$$
$$F = \beta\,\alpha\,\alpha\,\beta\,\alpha$$
$$G = \beta\,\alpha\,\alpha\,\alpha\,\beta$$
$$H = \alpha\,\beta\,\alpha\,\alpha\,\beta$$
$$I = \alpha\,\alpha\,\beta\,\alpha\,\beta$$
$$J = \alpha\,\alpha\,\alpha\,\beta\,\beta$$

	A	B	C	D	E	F	G	H	I	J	k_μ
$^2\Psi_1$	$\frac{1}{4}$	$\frac{3}{4}$	$\frac{1}{2}$	$\frac{1}{2}$	$1*$	-1	0	0	-1	1	4
$^2\Psi_2$	$\frac{1}{2}$	$\frac{1}{2}$	0	$1*$	$-\frac{1}{2}$	$-\frac{1}{2}$	0	-1	$\frac{1}{2}$	$\frac{1}{2}$	3
$^2\Psi_3$	$\frac{1}{2}$	$\frac{1}{2}$	$1*$	0	$-\frac{1}{2}$	$-\frac{1}{2}$	1	0	$-\frac{1}{2}$	$-\frac{1}{2}$	3
$^2\Psi_4$	$\frac{1}{3}$	$1*$	$-\frac{1}{2}$	$-\frac{1}{2}$	$-\frac{1}{4}$	$\frac{1}{4}$	$\frac{1}{2}$	$-\frac{1}{2}$	$-\frac{1}{4}$	$\frac{1}{4}$	$\frac{9}{4}$
$^2\Psi_5$	$1*$	$-\frac{1}{3}$	$-\frac{1}{3}$	$-\frac{1}{3}$	$\frac{1}{3}$	$-\frac{1}{3}$	$\frac{1}{3}$	$-\frac{1}{3}$	$\frac{1}{3}$	$-\frac{1}{3}$	2

BIBLIOGRAPHY

1. Coulson C.A., *Valence* 2nd edition, Clarendon Press, Oxford (1961).

2. Daudel R., Lefebvre R. and Moser C., *Quantum chemistry – methods and applications*, Interscience, New York (1959).

3. Pilar F.L., *Elementary quantum chemistry*, McGraw-Hill, New York (1968).

4. Parr R.G., *Quantum theory of molecular electronic structure*, W.A. Benjamin, New York (1963).

5. Eyring H., Walter J. and Kimball G.E., *Quantum chemistry*, John Wiley, New York (1944).

6. Clementi E., *I.B.M. J. Res. Dev.* 9, 2 (1965).

7. Clementi E., *J. chem. Phys.* 46, 3842 (1967).

8. Slater J.C., *Electronic structure of molecules*, McGraw-Hill, New York, (1963).

9. Cotton F.A., *Chemical applications of group theory*, Interscience, New York, (1963).

10. Shonland D., *Molecular Symmetry*, Van Nostrand, London and New York (1965).

11. Herzberg G., *Spectra of diatomic molecules*, Van Nostrand, Princeton N.J. (1950).

12. Nesbet R.K., *J. Math. Phys.* 2, 701 (1961).

13. Berthier G., in *Molecular orbitals in chemistry physics and biology*, ed. Löwdin P–O, and Pullman, B, Academic Press, New York (1964).

14. Roothaan C.C.J., *Rev. Mod. Phys.* 32, 179 (1960).

15. Nesbet R.K., *Proc. R. Soc.* A230, 312, 322 (1955).

16. Herzberg G., *Electronic spectra and electronic structure of polyatomic molecules*, Van Nostrand, Princeton (1966).

17. Boys S.F. , *Proc. R. Soc.* A200, 542 (1950).

18. Huzinaga S., *J. chem. Phys.* 42, 1293 (1965). Veillard A., *Theor. Chim. Acta* 12, 405 (1968). Basch H., Hornback C.J. and Moskowitz J.W., *J. chem. Phys.* 51, 1311 (1969).

19. Clementi E. and Davis D.R., *J. chem. Phys.* **45**, 2593 (1966).

20. Whitten J.L., *J. chem. Phys.* **44**, 359 (1966).

21. Barnett M.P. and Coulson C.A., *Phil. Trans. R. Soc.* **221**, 4243 (1951).

22. Shavitt I. and Karplus M., *J. chem. Phys.* **43**, 398 (1965).

23. Altmann S.L., in *Quantum Theory*, vol. 2, ed. Bates D.R., Academic Press, New York, (1962).

24. Clementi E., *J. chem. Phys.* **38**, 2248 (1963). *J. chem. Phys.* **39**, 175 (1963). *J. chem. Phys.* **42**, 2783 (1965).

25. Clementi E. and Veillard A., *J. chem. Phys.* **44**, 3050 (1966). Veillard A. and Clementi E., *J. chem. Phys.* **49**, 2415 (1968).

26. Clementi E., *J. chem. Phys.* **46**, 3851 (1967).

27. Goldfinger P. and Verhaegen G., *J. chem. Phys.* **50**, 1467 (1969).

28. Pople J.A., Santry D.P. and Segal G.A., *J. chem. Phys.* **43**, S129 (1965).

29. Pople J.A. and Segal G.A., *J. chem. Phys.* **44**, 3289 (1966).

30. Hoffmann R., *J. chem. Phys.* **39**, 1397 (1963).

31. Streitwieser A., *Molecular orbital theory for organic chemists*, John Wiley, New York and London (1961).

32. Roberts J.D., *Notes on molecular orbital calculations* W.A. Benjamin, New York (1961).

33. Dewar M.J.S., *The molecular orbital theory of organic chemistry*, McGraw Hill, New York (1969).

34. Richards W.G., Walker T.E.H., and Hinkley R.K., *Bibliography of Ab Initio molecular wave functions*, Clarendon Press, Oxford (1970).